NEMATOLOGY

LABORATORY INVESTIGATIONS

Morphologyand Taxonomy

JONATHAN D. EISENBACK

Library of Congress Cataloging-in Publication Data
Names: Eisenback, Jonathan D., 1952-
Title: Nematology Laboratory Investigations
Description: Mactode Publications [2022]
Identifiers: ISBN 978-1-893961-54-8 (softcover)
Subjects: LCSH: Nematodes, Nematology

Author and Editor: Jonathan D. Eisenback
Copy Editor: Marilyn L. Eisenback
Art Editor and Designer: Jonathan D. Eisenback

ISBN 978-1-893961-54-8

Printed in the United States
by
Kindle Direct Publishing
and Published by
Mactode Publications
Blacksburg, VA 24060

August 6/2022

This book is dedicated to my loving wife,
Marilyn
and my academic mother and the source of most of these investigations,
Hedwig Hirschmann.

John Halbrendt
is acknowledged for his encouragement
to complete this manual.

Preface

Laboratories are very necessary in introductory plant nematology courses. I believe that they are extremely valuable to the beginning student and well worth the effort that they require to teach. Laboratory experiences stimulate interest, arouse curiosity, and confirm information that is presented in lectures. Nematodes are extremely fascinating, and working with them in the laboratory is bound to excite all but the most uninitiated.

These investigations on the morphology and taxonomy of nematodes can be taught in one course or individually, according to the instructor's preference. Teachers are encouraged to select the exercises that suit the needs of their students, and students are encouraged to perform additional laboratories on their own initiative. I hope that these laboratories will accommodate many courses that are taught in a variety of academic settings. Labs have been selected to provide a basic introduction to plant nematology, as well as more advanced techniques.

This first volume of exercises covers the morphology and taxonomy of nematodes. More investigations that deal with many other aspects of nematology are currently being designed and tested in the classroom. They will be made available in a second volume.

I thank all who have contributed to this manual. In particular, I would like to thank Dr. Hedwig Hirschmann. She and her laboratory outlines were the inspiration for many of these laboratory exercises. In addition, I also thank those who trained me as a nematologist, including Drs. A. C. Triantaphyllou, Ken Barker, James Baldwin, Joseph Sasser, Donald Schmitt, and Dick Powell at North Carolina State University. Many of these labs were taught to me by these outstanding nematologists and teachers.

J.D.E.

Contents

Investigation 1

Preparation of Temporary and Semi-Permanent Mounts

Nematodes are small, colorless roundworms that are most often observed with a microscope. To be seen clearly, they are mounted on a glass slide and covered with a coverslip. In order to see all of the minute details, the mounts are examined with an oil immersion objective lens. Usually, several adult specimens of similar appearance in the dissecting microscope are selected and mounted for identification to species. The specimens are preserved for a few hours to indefinitely, and, if their structures are clear and clearly displayed, it may be possible to make an accurate identification. The condition of the specimen, the selection of the mounting medium, and the orientation of the nematodes are important because morphological details are easily obscured by improper preparation techniques. Mounts may be temporary, semi-permanent, or permanent.

The purpose of this lab is to demonstrate techniques to make temporary and semi-permanent mounts of nematodes. Temporary mounts are made with the specimens mounted in water. They are observed immediately because they last only for a few hours or less. Semi-permanent mounts preserve specimens for a few hours to several days. Permanent mounts can be observed for many years, but they are more difficult to prepare than temporary and semi-permanent mounts. (See how to make permanent mounts in Investigation 22.)

Objectives of the Laboratory

Upon completion of this lab the student should be able to:

1. **Prepare** ringed slides for mounting nematodes.
2. **Make** a temporary mount of a nematode.
3. **Outline** the procedure for making semi-permanent mounts.
4. **Collect** nematodes from various habitats.
5. **Extract** nematodes from soil and plant tissues utilizing a Baermann funnel.

Materials needed for this laboratory:

Nematode species:

Rhabditis sp. or other suitable nematode

For each student:

Forceps
Light microscope
Nematode pick
Slide storage box

For each pair of students:

Alcohol lamp
Baermann funnels
Butyl acetate
Cheesecloth
2% formalin
Glass coverslips
Glass microscope slides
Immersion oil
Matches
Nail polish
Plastic bags
Slide labels
Slide ringer
Small paint brush
Syracuse dishes
Trowels or soil sampling tube
Water

Work Sheet

I. Practice making ringed slides for mounting nematodes.

A. Wipe a pre-cleaned slide with a clean cheesecloth. (Do not clean the slide with alcohol because the nail polish will not stick to the glass properly.)

B. Center a glass slide on the slide ringer.

C. Dip a small brush into a thinned solution of nail polish.

D. Gently spin the slide in the ringer and touch the loaded brush to the slide using the scribed circle as a guide. (Angle the brush so that the nail polish is dragged off the brush onto the turning slide.)

E. Vary the height of the ring to match the different diameters of various nematodes.

F. Allow the ring to dry for at least 30 minutes in a level slide tray before mounting a specimen. (Keep it covered to prevent dust from ruining the ring.)

1. Why is it necessary to support the coverslip? ______________________________

____

____

II. Practice making temporary mounts in water.

A. Pick 10 specimens of each species with a nematode pick and transfer them into a drop of water on a heavily ringed slide.

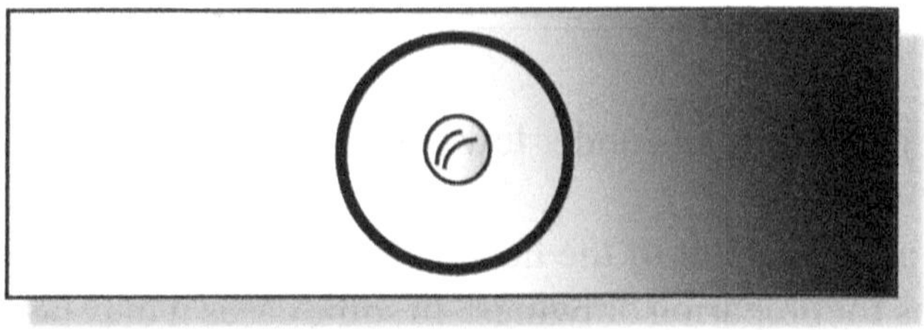

B. Gently kill the nematodes over an alcohol flame. Use just enough heat to straighten the specimens.

C. Gently place a coverslip on top of the drop, and carefully absorb the excess water with triangular pieces of filter paper. (Placing the coverslip is the hardest part in making mounts of nematodes.)

D. Seal the slide with a thick layer of nail polish. (Be sure to use enough so that the water will not evaporate.)

III. Practice making semi-permanent mounts in 2% formalin.

A. Pick 10-15 specimens with a nematode pick and transfer them into a drop of water on a heavily ringed slide.

B. Gently kill the nematodes over an alcohol flame. Use just enough heat to straighten the specimens.

C. Transfer 5-7 of the best specimens to the bottom of a drop of 2% formalin on a ringed slide.

D. Gently place a coverslip on top of the drop and carefully absorb the excess formalin with triangular pieces of filter paper.

E. Seal the slide with a thick layer of nail polish. (Be sure to use enough so that the formalin will not evaporate.)

F. Label the slide with the name of the genus and species (if known), the number and sex of specimens, and the date that the slide was made.

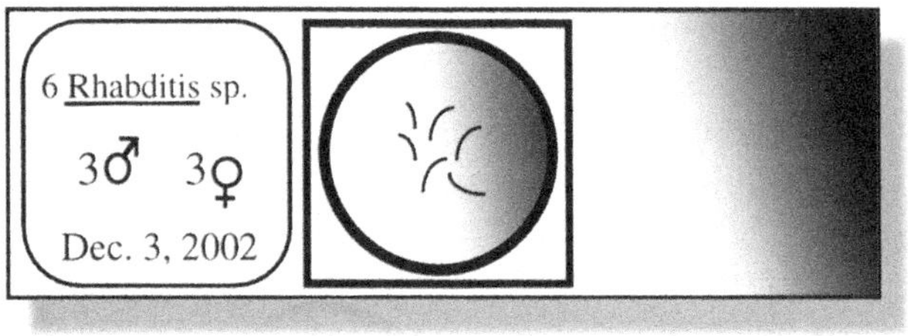

G. Observe the specimens under low and high power on the compound light microscope.

H. Make drawings of all of the details that you can see. The instructor will give general directions for preparing drawings. In some cases it may be necessary to make composite drawings to show all of the structures that were seen in various specimens.

2. How can you recognize the position in which your specimens are lying? ____________

____

____

3. In what position are most of your specimens lying? ____________________

____

IV. Prepare a collection of semi-permanent mounts (20 slides).

A. Collect several soil samples from different habitats. (Keep the samples in closed plastic bags to prevent drying.)

B. Process the samples as soon as possible after collecting by placing 50 cc of soil on a Baermann funnel. (Processing equipment will be available in the lab.)

C. Draw off the funnel 24 hours after processing and store the nematodes in a Syracuse watch dish in the refrigerator. (Stack the dishes, provide them with a cover to prevent them from drying out, and label them with your name.)

D. Prepare semi-permanent mounts of the extracted nematodes in 2% formalin as outlined on pages 3 and 4.

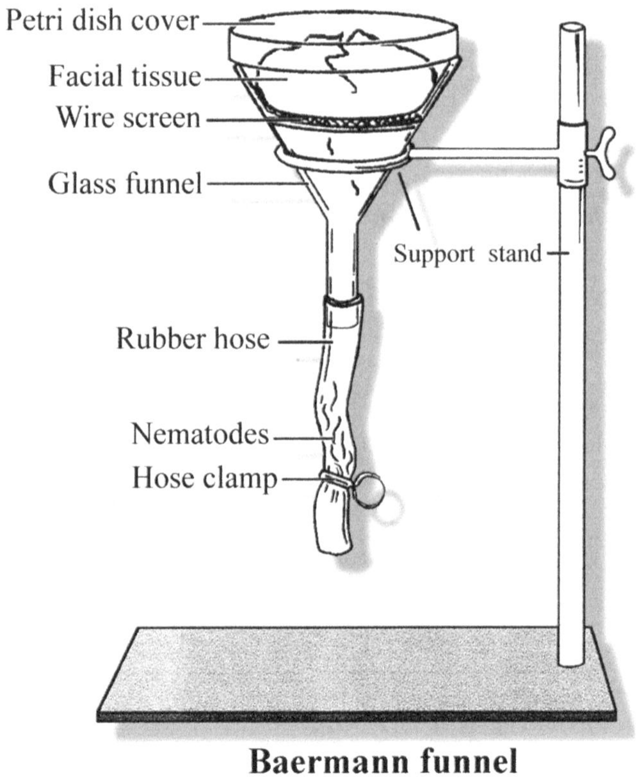

Baermann funnel

E. Identify the nematodes to genus. (Later labs will provide the necessary information and keys for identifying the plant-parasitic nematodes to genus.)

F. Each mount should contain at least 5 adult nematodes of the same genus. Make the mounts carefully; preparations containing air bubbles are useless because specimens float, dry out quickly, and cannot be observed with oil immersion. The quality of slides will be evaluated by the instructor.

G. Label the slides with the following information:

Label on left side: Genus, number of specimens, and sex.
Label on right side: Your name, source of the specimens, locality, and date prepared.

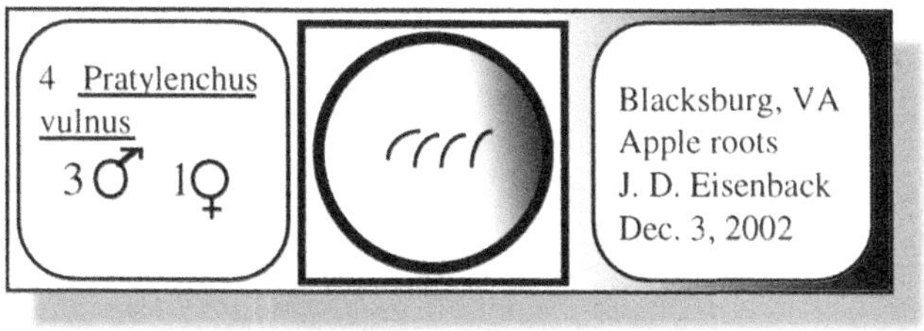

H. Turn in slides of at least 20 different genera as soon as they are made because the specimens deteriorate rapidly in semi-permanent mounts.

4. What considerations should be taken into account in the collection and storage of soil and plant tissues that will be used for nematode extraction? ______________________

____

____

____

5. How does a Baermann funnel work? ______________________________________

____

____

____

6. Will a Baermann funnel work equally well for all nematodes? Why or why not? ______

____

____

____

Investigation 2

Gross Morphology of Free-Living Nematodes

Free-living nematodes are common inhabitants of the soil environment. Millions of individuals can live in the top 20 centimeters of one square meter of soil; they feed on bacteria, fungi, and dead organic matter. The morphology of the body wall, as well as the internal anatomy, is easily observed in whole specimens, and even live specimens can be examined. In order to see the minute details of nematode morphology, it is necessary to obtain specimens in good physiological condition, to prepare them with a suitable technique for observation, to adjust the microscope for maximum resolution, to compare the morphology of several specimens, and to look and look and look.

The morphology of free-living nematodes reflects their physiological function. Usually, the buccal cavity is pronounced and the esophagus is muscular, either throughout its entire length or in specialized areas for pumping. Esophageal glands may be difficult to detect, especially if they are interspersed within the muscular esophagus. (You may want to compare the morphology of free-living nematodes with that of plant-parasitic nematodes in Investigation 3.)

Objectives of the Laboratory

Upon completion of this lab the student should be able to:

1. **Observe** the gross morphology of a free-living nematode.

2. **Draw** the major body systems of the female and male of a free-living nematode.

Materials needed for this laboratory:

Nematode species:

Rhabditis sp. or other suitable nematode

For each student:

Forceps	Nematode pick
Light microscope	Slide storage box

For each pair of students:

Alcohol lamp	Matches
Butyl acetate	Nail polish
2% formalin	Slide labels
Glass coverslips	Slide ringer
Glass microscope slides	Small paint brush
Immersion oil	Water

Work Sheet

I. Make semi-permanent slides of females and males of *Rhabditis* sp.

A. Pick 10-15 females and 10-15 males of *Rhabditis* sp. and transfer them into a drop of water on a ringed slide.

B. Kill the nematodes gently over the alcohol flame. Use just enough heat to straighten the specimens.

C. Transfer 5-7 of the best specimens to the bottom of a drop of 2% formalin on a ringed slide.

D. Gently place a coverslip on top of the drop and carefully absorb the excess formalin with triangular pieces of filter paper. (Placing the coverslip is the hardest part in making mounts of nematodes.)

E. Seal the slide with a thick layer of nail polish. (Be sure to use enough so that the formalin will not evaporate.)

F. Prepare 1-2 slides each of females and males.

*1. What are the gross differences between female and male nematodes?*______________

____

____

____

II. Study the gross morphology of the *Rhabditis* females and males.

A. Study the gross morphology of female and male nematodes under low power. Compare and contrast the differences between the sexes. (Use low power magnification.)

B. Study the gross morphology of the anterior end of the female and the tails of the female and male. (Use high power magnification.)

C. Make drawings of all of the details that you can see. The instructor may give additional directions for preparing drawings. In some cases, it may be necessary to make composite drawings to show all of the structures that are visible. Draw just what you see! Do not copy the drawing in this investigation or from other sources. The instructor can ensure that you are correctly seeing and interpreting the specimens by looking at your drawings. Also, please make your drawings large enough to include all of the minuscule details.

III. Make the following drawings:

A. Entire nematode: female and male (low power).

B. Anterior portion of female including esophagus (high power).

C. Tail of female (high power).

D. Tail of male (high power).

2. *Based on the morphology that you observed, do you think that* Rhabditis *is a plant parasite or a free-living nematode?* ______________________________

Why? ____

____

3. *Compare and contrast the differences between the anterior ends of the female and male.* ____

____

____

____

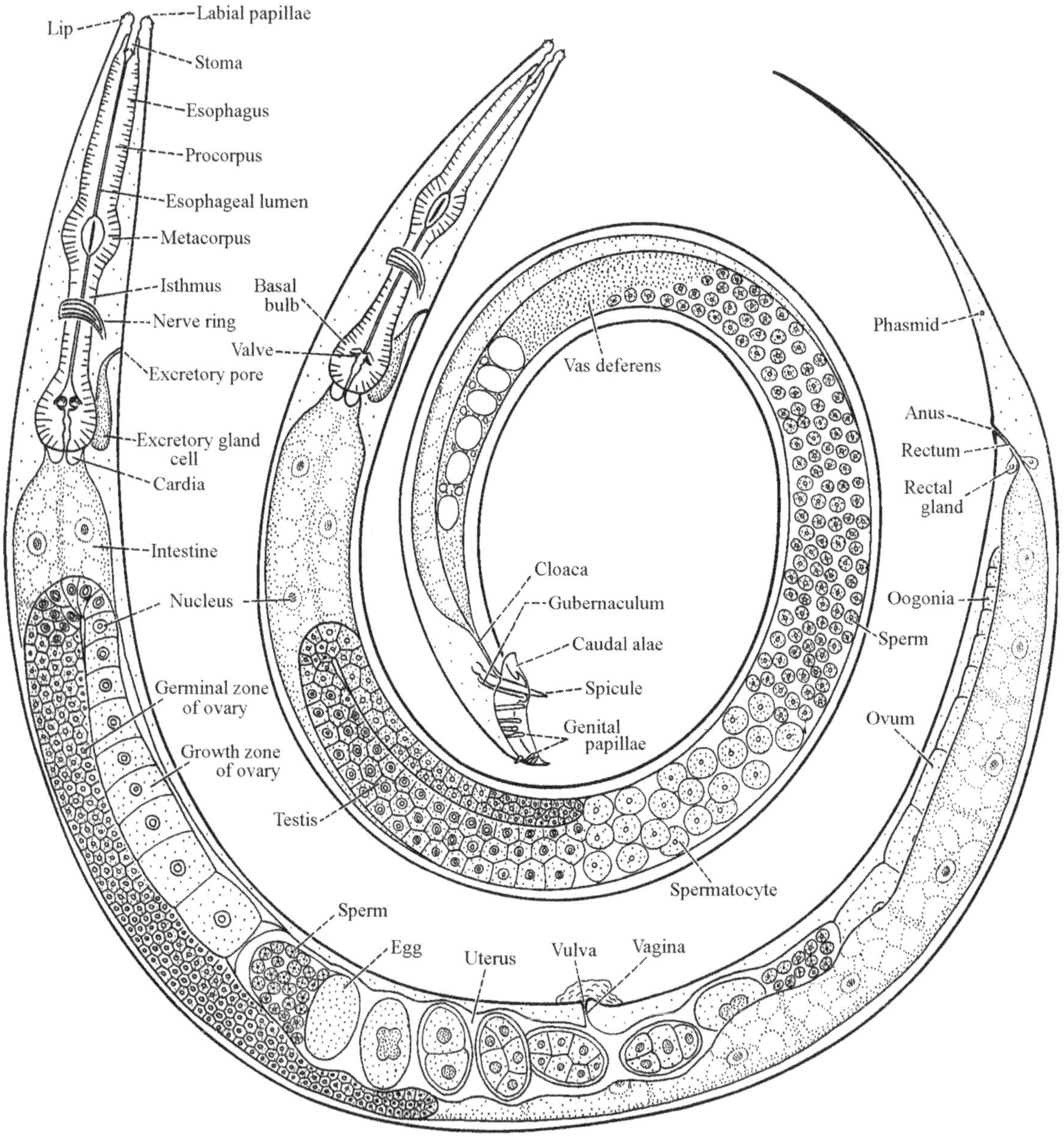

Fig. 2.1 Gross morphology of a typical soil-borne free-living nematode, female and male (after Hirschmann).

4. *Compare and contrast the differences between the tails of the female and male.* _____

____

____

____

Investigation 3

Gross Morphology of Plant-Parasitic Nematodes

All plant-parasitic nematodes possess a protusible hypodermic needle-like stylet that they use to feed on plants. It is usually hollow with few exceptions (*Trichodorus* and *Paratrichodorus*). Most plant-parasitic nematodes are vermiform for their entire life, but juveniles and females of some species become swollen after they begin feeding on plant tissues. These saccate females are sedentary parasites, and they sacrifice their mobility to increase their reproductive capacity. Males of these species remain vermiform and, in most instances, do not feed as adults.

Usually, the esophagus of plant-parasitic nematodes is less muscular than that of the free-living nematodes. Often the pumping portion is limited to the median bulb; however, the esophageal glands are usually very prominent in plant-parasitic nematodes because they produce the secretions that are necessary for establishing and maintaining the parasitic relationship. (You may want to compare the morphology of plant-parasitic nematodes with that of free-living nematodes in Investigation 2.)

Objectives of the Laboratory

Upon completion of this lab the student should be able to:

1. **Observe** the gross morphology of a plant-parasitic nematode.
2. **Draw** the major body systems of a plant-parasitic nematode.
3. **Identify** swollen females of sedentary endoparasitic nematodes.

Materials needed for this laboratory:

Nematode species:

Meloidogyne sp. *Tylenchorhynchus* sp.

For each student:

Forceps
Light microscope
Nematode pick
Slide storage box
Surgical eye knife

For each pair of students:

Alcohol lamp
BPI dish
Butyl acetate
Dissecting needles (2)
2% formalin
Glass coverslips
Glass microscope slides
Immersion oil
Matches
Nail polish
Pipettes
Slide label
Slide ringer
Small paint brush
Syracuse dish
Water

Work Sheet

I. Make semi-permanent slides of females and males of *Tylenchorhynchus* sp. (*Tylenchorhynchus* sp. females and males are representatives of the vermiform type.)

A. Pick up 20 females and 20 males with the pulp canal file and transfer them to a drop of water on ringed slides.

B. Note the shape, size, color and movement of the specimens.

C. Kill the nematodes by gentle heat over an alcohol flame.

D. Transfer each of 6-8 females or males to a drop of 2% formalin on ringed slides and make them into semi-permanent mounts. Prepare 2-3 slides. Seal the mounts with a thick ring of nail polish.

E. Study the gross morphology of the *Tylenchorhynchus* sp. females and males.

F. Make drawings of all of the details that you can see. The instructor will give general directions for preparing drawings. In some cases, it may be necessary to make composite drawings to show all of the structures that are visible. Draw just what you see! Do not copy the drawing in this investigation or from other sources. The instructor can insure that you are correctly seeing and interpreting the specimens by looking at your drawings. Also, make your drawings large enough so that you can include all of the minuscule details.

II. Make the following drawings:

A. Entire nematode: female and male (low power).

B. Anterior portion of female including esophagus (oil immersion).

C. Tail of female (oil immersion).

D. Tail of male (oil immersion).

1. Describe the morphology of the stylet of Tylenchorhynchus *sp. females and males.*

____

____

____

2. How many parts of the esophagus can you see in Tylenchorhynchus *sp. ?* __________

____

____

3. Where are the openings of the esophageal glands located in the esophagus of Tylenchorhynchus *sp. ?* ______________________________

____

4. Compare and contrast the tails of females and males of Tylenchorhynchus *sp.* ______

____

____

____

____

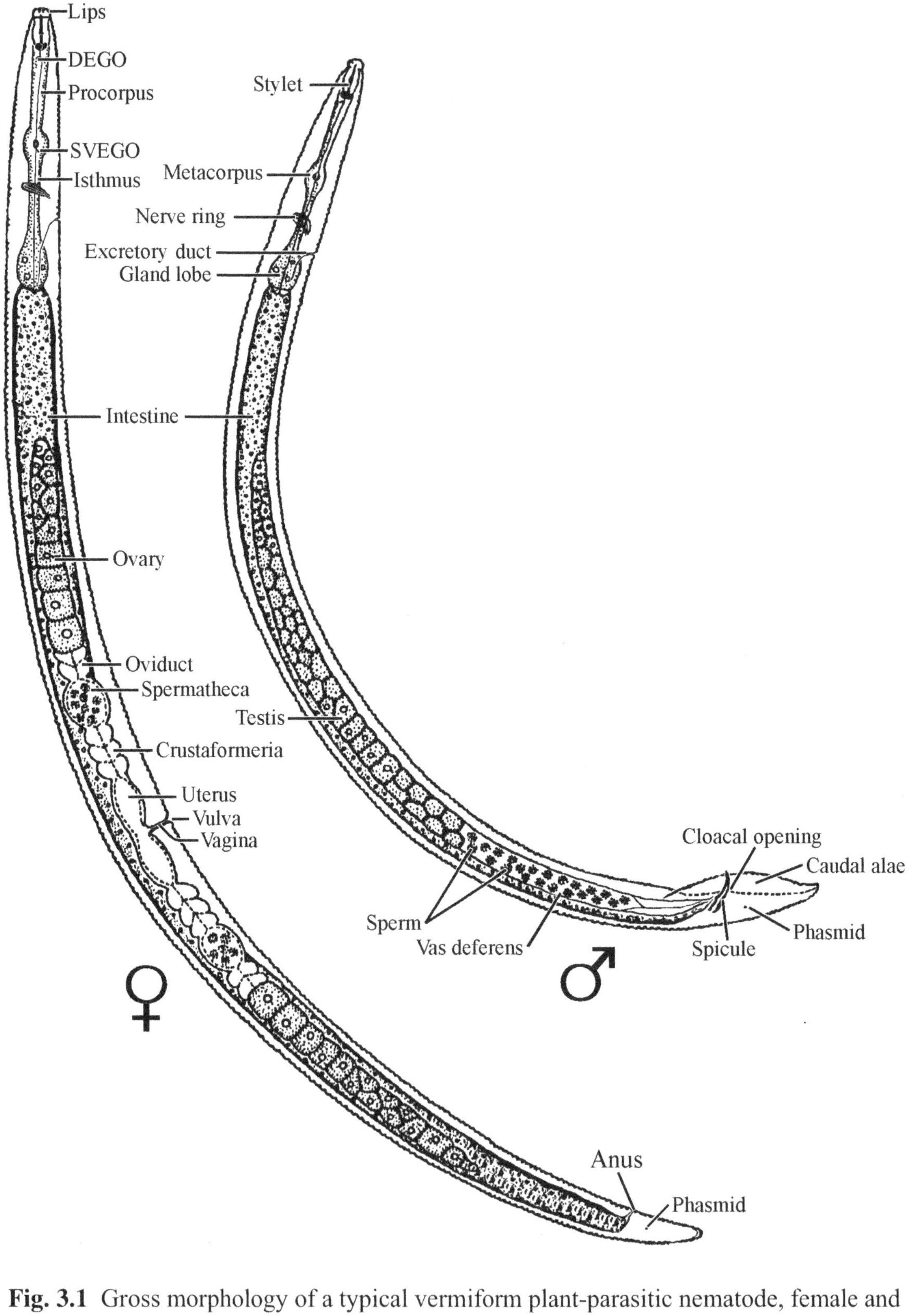

Fig. 3.1 Gross morphology of a typical vermiform plant-parasitic nematode, female and male (after Hirschmann).

III. Examine the morphology of females and males of *Meloidogyne* sp. (*Meloidogyne* sp. females are representatives of the saccate type of plant-parasitic nematodes.)

A. Examine the provided root material infected with *Meloidogyne* sp. Observe the galled appearance of the roots and the brownish egg masses on the gall surface.

B. Tease out some *Meloidogyne* sp. females present in the galls using dissecting needles. Transfer 10 females with a pipette to a BPI dish containing 2% formalin.

C. Pick out 3-4 *Meloidogyne* sp. males and transfer them to a drop of water on a slide. Kill them by gentle heat and make a semi-permanent formalin mount.

D. Examine the females and males at low power on the microscope and make outline drawings of them.

E. Make drawings of all of the details that you can see. The instructor will give general directions for preparing drawings. In some cases, it may be necessary to make composite drawings to show all of the structures that are visible. Draw just what you see! Do not copy the drawing in this investigation or from other sources. The instructor can insure that you are correctly seeing and interpreting the specimens by looking at your drawings. Also, make your drawings large enough so that you can include all of the minuscule details.

IV. Make the following observations and drawings:

A. Entire nematode: female and male (low power).

B. Observe the sexual dimorphism in *Meloidogyne*: Females are pyriform, males are fusiform.

5. *What is the significance of the swelling of the females of* Meloidogyne *sp. ?* ________

____

____

6. *Compare and contrast the body shape of females and males of* Meloidogyne *sp.* ____

____

____

____

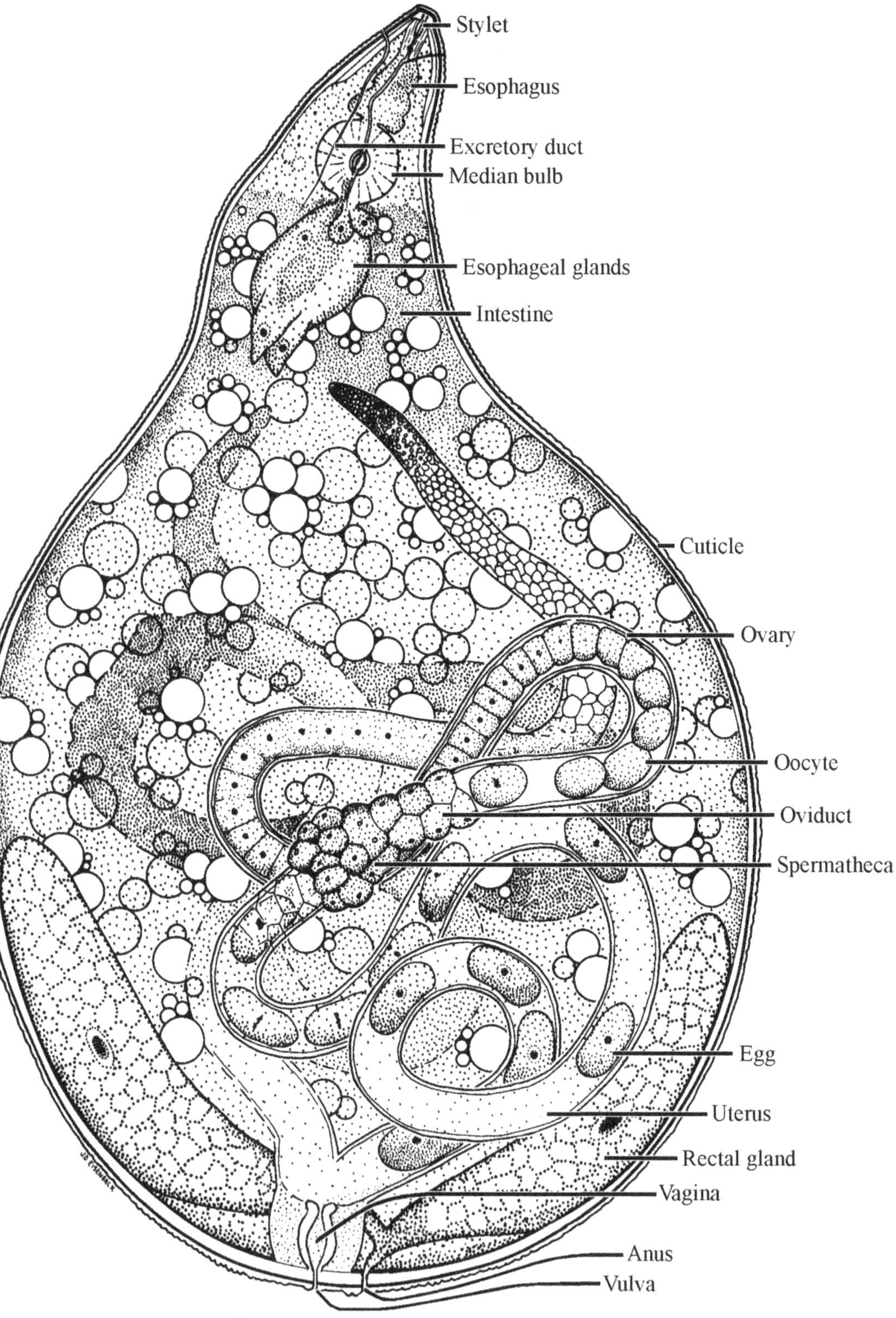

Fig. 3.2 Gross morphology of a typical pyriform plant-parasitic nematode (after Eisenback).

7. *Explain the trade-off between the loss of mobility and the increase in body size in females of* Meloidogyne *spp.* ______________________________

____

____

____

8. *Compare and contrast the differences in the morphology of females of* Tylenchorhynchus *sp. and* Meloidogyne *sp.* ______________________________

____

____

____

____

Investigation 4

Cuticular Markings

Nematodes are decorated with various types of cuticular markings that are useful for identification. These markings are either ornamental or sensory; however, ornamental does not imply lack of function. Cuticular markings are generally external, although a few markings occur below the surface, and may be considered as important morphological characters that are useful for the description and subsequent identification of nematode species. These markings are sometimes difficult to see and can be enhanced with proper lighting techniques on the microscope.

For some characters, there is a progression from one state to the next. For example, transverse markings vary from none, to slightly indented, to moderately deep, to very deep. These different character states are not clearly defined, nor are they distinctly separated from each other. Therefore, much experience may be necessary to acquire the ability to accurately identify a particular character state.

Objectives of the Laboratory

Upon completion of this lab the student should be able to:

1. **Draw** the progression of transverse markings from striae to apparent segmentation.
2. **Study** the various types of longitudinal markings.
3. **Observe** subcuticular punctations.
4. **Examine** the lips and their modifications.

Materials needed for this laboratory:

Nematode species:

Acrobeles sp.
Aphelenchus sp
Heterodera sp
Hoplolaimus sp
Mesocriconema sp.
Oesophagostomum sp.
Pratylenchus sp.
Rhabditis sp.
Tylenchorhynchus sp.
Marine nematode with punctations

For each pair of students:

- Alcohol lamp
- Butyl acetate
- 2% formalin
- Glass coverslips
- Glass microscope slides
- Immersion oil
- Matches
- Nail polish
- Slide labels
- Slide ringer
- Small paint brush
- Water

Work Sheet

I. Examine the progression of transverse markings from striae to apparent segmentations and make drawings of the following:

A. Striation: *Pratylenchus* sp.

B. Annulation: *Hoplolaimus* sp.

C. Apparent segmentation: *Mesoriconema* sp.

D. Modification of annulation: *Heterodera* sp. Make a sketch of the zig-zag pattern of the cyst wall. Also notice the punctations in the inner layer of the cyst wall.

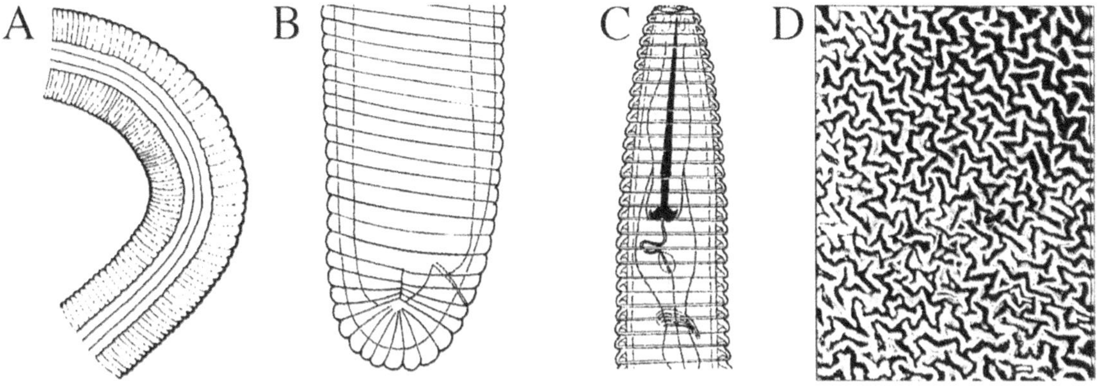

Fig. 4.1. Progression of transverse body markings from striae to apparent segmentation and other variations. A. Striae in *Pratylenchus* sp. (after Corbet) B. Annulation in *Hoplolaimus* sp. (after Sher) C. Apparent segmentation in *Mesocriconema* sp. (after Orton Williams) D. Zig-zag like pattern in the body wall of cyst nematodes, *Globodera* and *Heterodera* spp. (after Hirschmann)

1. Explain the differences between striation, annulation, and apparent segmentation. ___

II. Study the various types of longitudinal markings and make drawings of the following:

A. Study the longitudinal ridges and annules that combine to form tesselation in *Tylenchorhynchus* sp.

B. Examine the longitudinal alae in the following specimens:
   1. Four lines occur in the lateral field of *Tylenchorhynchus* sp.
   2. Ten lines are present in the lateral field of *Aphelenchus* sp.
   3. Observe the cross-section of the lateral field of *Hoplolaimus* sp.

C. Draw the tail of a *Hoplolaimus* sp. male and note the difference between the morphology of the bursa (caudal alae) of *Hoplolaimus* sp. and *Rhabditis* sp.

D. Observe the cervical alae of *Oesophagostomum* sp.

2. Compare and contrast the morphology of the mail tail of Hoplolaimus *and* Rhabditis *spp.* ___

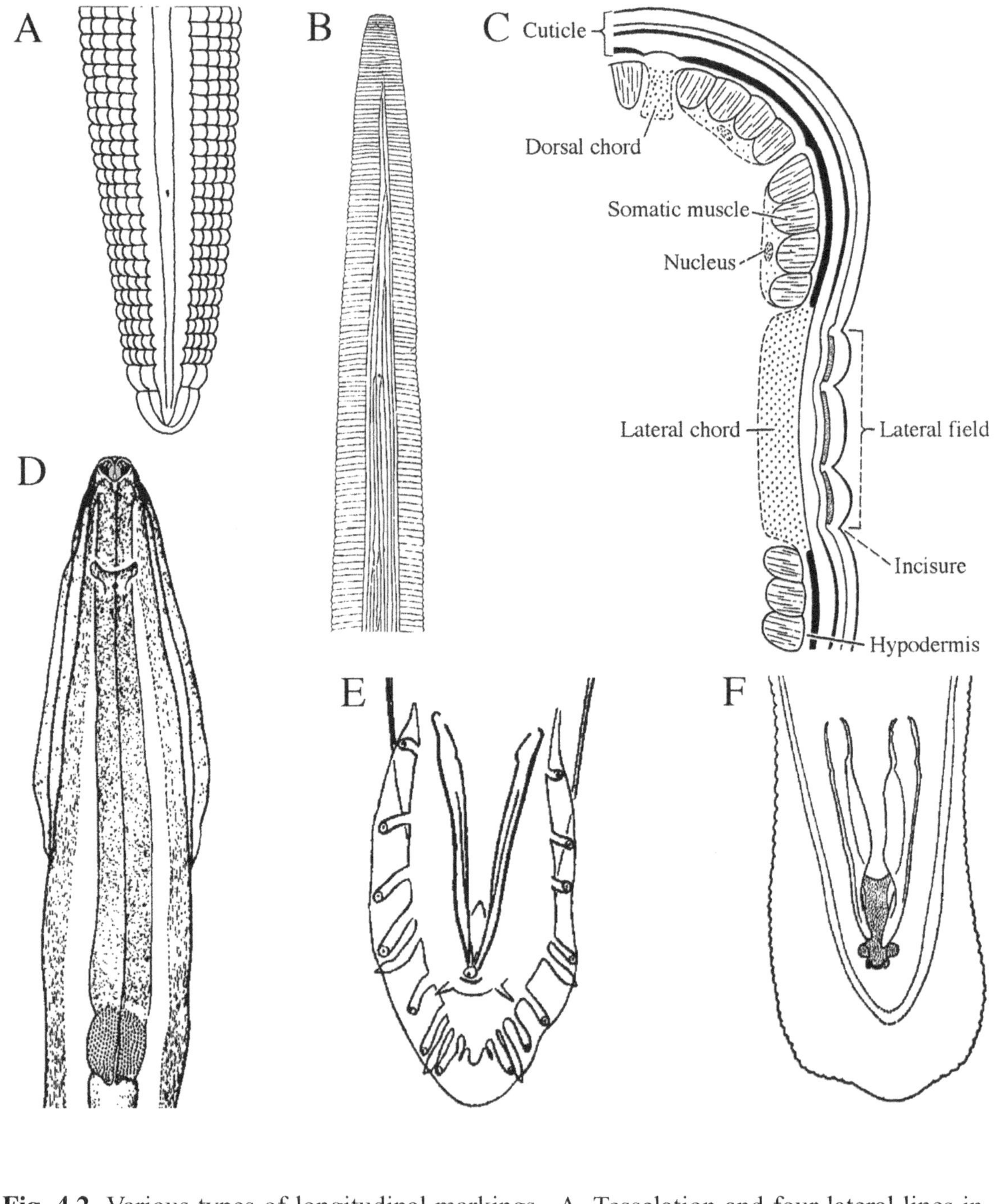

Fig. 4.2 Various types of longitudinal markings. A. Tesselation and four lateral lines in *Tylenchorhynchus* sp. B. Ten lateral incisures in *Aphelenchus* sp. (after Goodey & Hooper) C. Cross-section of the body wall of *Hoplolaimus* sp. D. Cervical alae of *Oesophagostomum* sp. E. Tail of *Rhabditis* sp. (lateral view) F. Tail of *Hoplolaimus* sp. (lateral view) (C-F After Hirschmann)

III. Observe and draw the punctations in the cuticle of a marine nematode.

IV. Examine the lips and their modifications and make drawings of the following:

A. Study the lips and their papillae of *Rhabditis* sp. in side view.

B. Note the labial probolae which are modifications of the lips in *Acrobeles* sp.

C. Observe the cephalic framework in *Hoplolaimus* sp. and *Pratylenchus* sp.

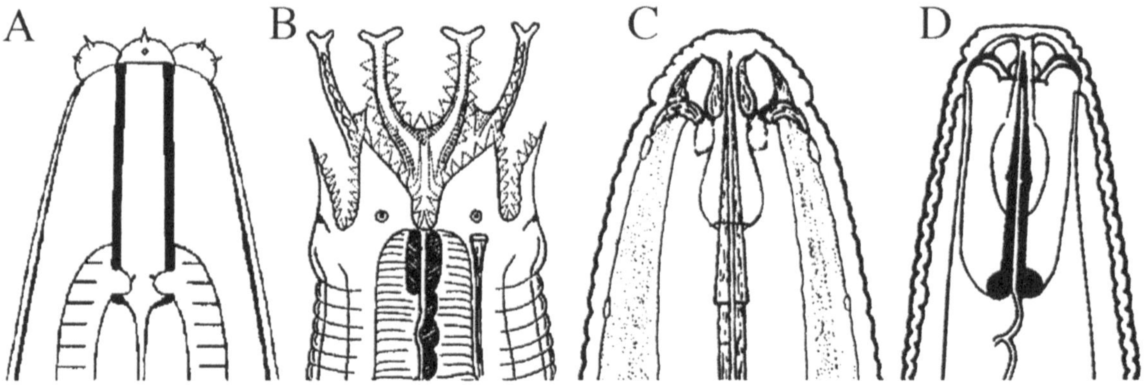

Fig. 4.3 Lips and their modifications. A. Lips of *Rhabditis* sp. (after Hirschmann) B. Probalae of *Acrobeles* sp. (after Thorne) C. Heavy cephalic framework of *Hoplolaimus* sp. (after Sher) D. Moderate framework of *Pratylenchus* sp. (after Corbett)

3. What is the purpose of the cephalic framework in plant-parasitic nematodes? ______

____

____

____

____

____

____

Investigation 5

Cuticular Layering

The cuticle of most nematodes is often arranged into four fundamental layers: epicuticle, exocuticle, mesocuticle, and endocuticle. Both the external and internal cuticular structures are important taxonomic characters. In the root-knot nematodes, *Meloidogyne* spp., the cuticle surrounding the vulva and anus forms a unique fingerprint-like pattern that is useful for the identification of the four most common species. Preparing these patterns for observation is difficult and tedious but not impossible.

Objectives of the Laboratory

Upon completion of this lab the student should be able to:

1. **Draw** the cuticular layering of an animal-parasitic nematode.
2. **Describe** the cuticular layering of a plant-parasitic nematode.
3. **Make** a perineal pattern of a *Meloidogyne* spp.
4. **Compare** the perineal patterns of the four most common *Meloidogyne* species.

Materials needed for this laboratory:

Nematode species:

Ascaris lumbricoides (cross-sections on prepared slides)
Heterodera trifolii or *H. glycines*
Meloidogyne arenaria | *M. hapla*
M. incognita | *M. javanica*

For each student:

Dissecting microscope | Nematode pick
Forceps | Light microscope
Slide storage box

For each pair of students:

Dissecting needles	45% lactic acid
2% formalin	Nail polish
Glass microscope slides	Plastic petri dish
Glass coverslips	Slide labels
Glycerin	Small eye knife or single-edged razor blade
Half spear	Syracuse dish
Immersion oil	Water

Work Sheet

I. Examine the cuticular layering of an animal-parasitic nematode (*Ascaris lumbricoides*).

A. Observe a transverse section of the cuticle under high power.

B. Make drawings of a close-up view of a transverse section of the cuticle (use oil immersion).

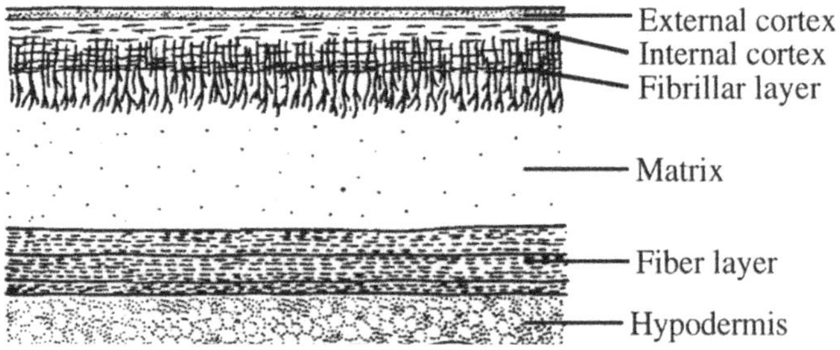

Fig. 5.1 Transverse section of cuticle of *Ascaris lumbricoides* (After Hirschmann, 1960).

II. Observe the cuticular layering of a plant-parasitic nematode (*Heterodera trifolii*).

A. Transfer several brown cysts to a Syracuse watch glass filled with 2% formalin.

B. Select individual brown cysts and make thin transverse sections of the cyst wall with the eye knife in a drop of lactophenol on a plastic dish.

C. Make lactophenol mounts and observe structure of cyst wall.

D. Make a drawing of a close-up view of a transverse section of the cuticle (use oil immersion).

III. Make a perineal pattern of a *Meloidogyne* sp.

A. Select galls with mature females and place in Syracuse dish with tap water. (Single galls are preferable to compound galls.) Pick out 6-8 adult females by teasing the root tissue apart with forceps and half spear.

A

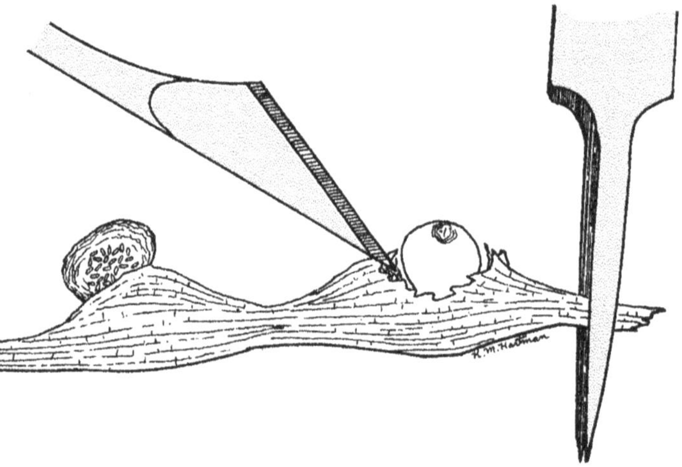

B. Cut off the neck of the female with the eye knife and gently push out the body contents with the forceps.

C. Place the cuticle in a drop of 45% lactic acid in a plastic Petri dish. Collect 6 to 8 cuticles in the drop and let them stand for 15 - 30 minutes before cutting.

B

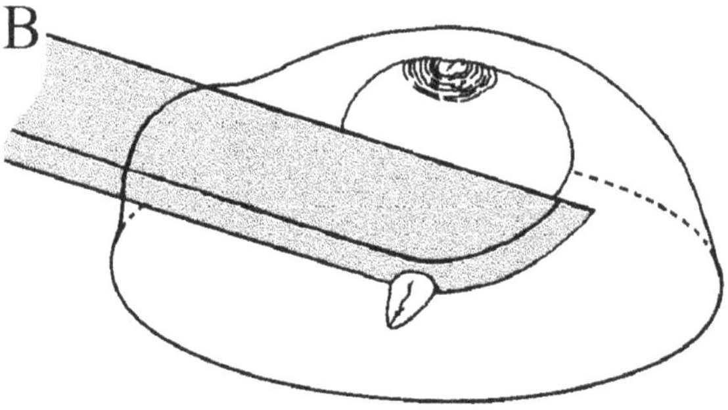

C

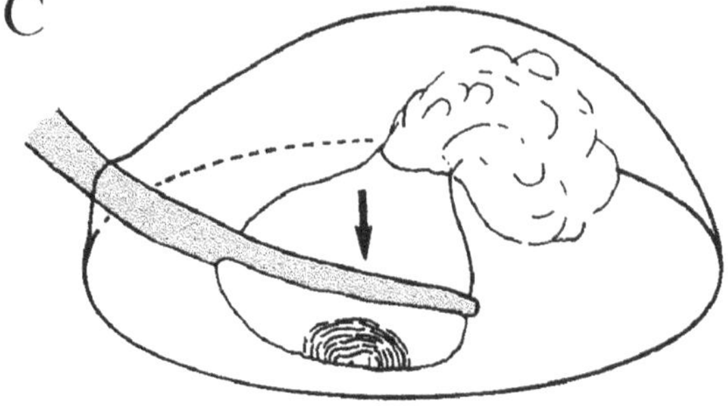

D. Cut the cuticle in half (equatorially) with an eye knife in a drop of 45% lactic acid.

E. Trim the cuticle with the perineal pattern so that it is square. Cut 6 to 8 patterns. (Cut only on the surface of a plastic Petri dish to prevent dulling the cutting edge.)

D

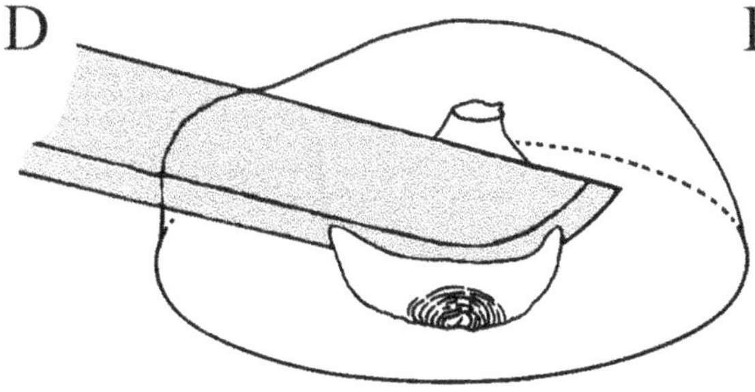

E

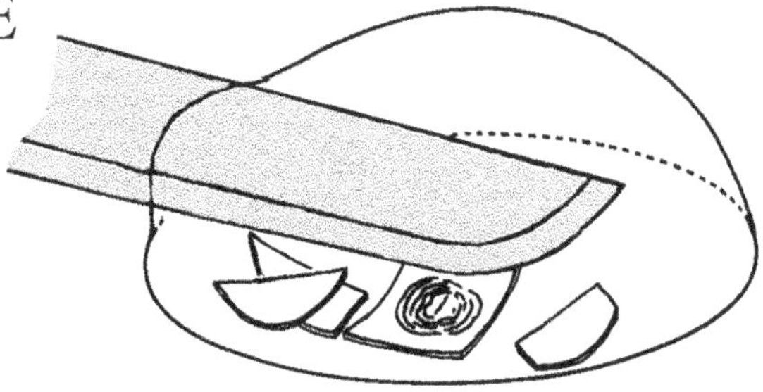

F. Thoroughly clean debris from the perineal pattern with a nematode pick.

G. Transfer the perineal pattern with a nematode pick to a drop of glycerin on a clean glass microscope slide. Align the perineal patterns so that they are in a straight line and the anus is oriented down. Place the interior surface of the cuticle against the glass and press the pattern gently against the glass with a nematode pick.

F

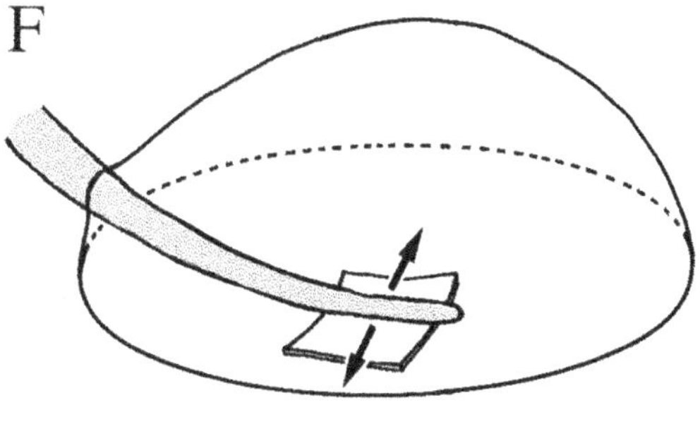

G

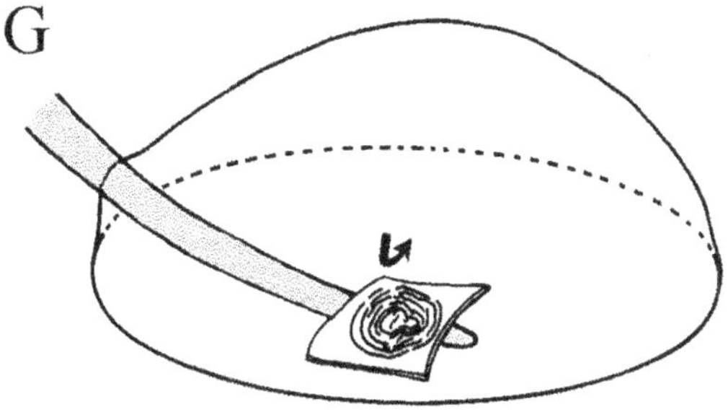

H. Gently place the coverslip in a drop of glycerin. Absorb excess glycerin with a small triangular piece of filter paper, or additional glycerin can be added by placing a small drop on the edge of the coverslip.

I. Seal the coverslip with nail polish and label the slide.

I

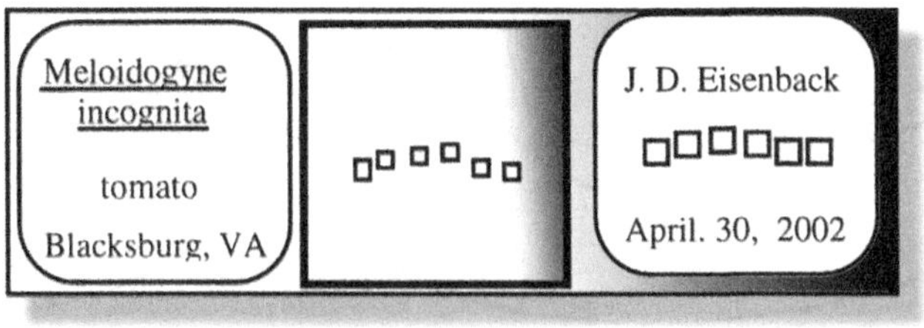

IV. Observe and draw the perineal patterns of the four most common *Meloidogyne* spp. (Fig. 5.2)

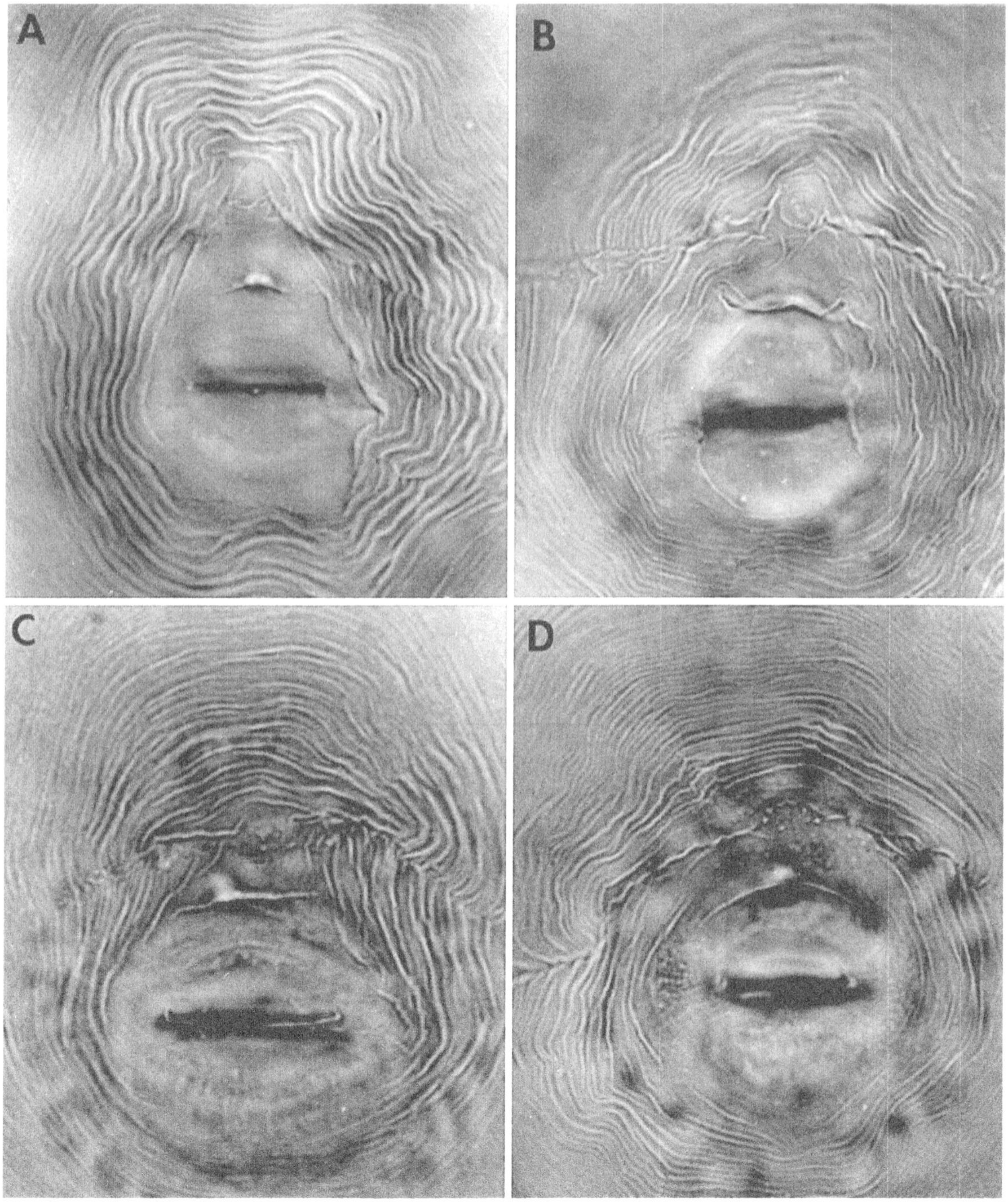

Fig. 5.2 Perineal patterns of the four most common *Meloidogyne* species. A. *M. incognita* B. *M. javanica* C. *M. arenaria* D. *M. hapla* (after Eisenback)

1. *Describe the major identifying characters for each of the four common species of* Meloidogyne.

M. incognita ______________________________

____

____

____

M. javanica ______________________________

____

____

____

M. arenaria ______________________________

____

____

____

M. hapla ______________________________

____

____

____

Investigation 6

Body Wall

Nematodes are multicellular animals (metazoans) with true tissues (umetazoans) that have a true mesoderm with a persistent blastocoel (= pseudocoel) between the gut and body wall. They are triploblastic, having three germcell layers including ectoderm (cuticle and hypodermis), mesoderm (somatic muscles), and endoderm (intestine). The body plan of nematodes consists of a tube within a tube. The body wall forms the outer tube, and the digestive system forms the inner tube. Animals with a similar body plan include other psuedocoelmates such as the Nematomorpha (horsehair worms), Entoprocta (entoprocts), Rotifera (rotifers), Gastrotricha (gastrotrichs), and Kinorhyncha (kinorhynchs).

Objectives of the Laboratory

Upon completion of this lab the student should be able to:

1. **Sketch** the cross-section of the espohagus of *Ascaris lumbricoides*.

2. **Draw** the generalized gross morphology of the cross-section of a female *A. lumbricoides*.

3. **Illustrate** a somatic muscle cell, a section of the intestinal wall, and various parts of the female and male reproductive systems of *A. lumbricoides*.

Materials needed for this laboratory:

Nematode species:

Prepared cross-sections of *Ascaris lumbricoides*:

Esophagus
Female reproductive system
Male reproductive system

For each student:

Light microscope
Slide storage box

For each pair of students:

Immersion oil

Work Sheet

I. Examine cross-sections of *Ascaris lumbricoides* with a microscope, studying the body wall and internal organs.

A. Make a sketch of the esophagus cross-section of *Ascaris*. Pay particular attention to the well-developed hypodermis.

B. Study the gross morphology of the cross-section of the *Ascaris lumbricoides* female and draw a general view under the low power of the microscope.

C. Draw close-up views of the following using the high power of the microscope:
1. Two or three of the coelomyarian muscle cells.
2. Section of the intestinal wall.
3. Growth zone of the ovary.
4. Uterus with eggs.

D. Study the gross morphology of the cross-section of the *Ascaris lumbricoides* male and draw a general view under the low power of the microscope.

E. Make a close-up sketch of the vas deferens with spermatids.

1. How many somatic muscle cells are contained in one cross-section of Ascaris lumbricoides *at the level of the esophagus?* ______________________

____

2. Describe the innervation processes of the somatic muscles. ______________

____

____

____

____

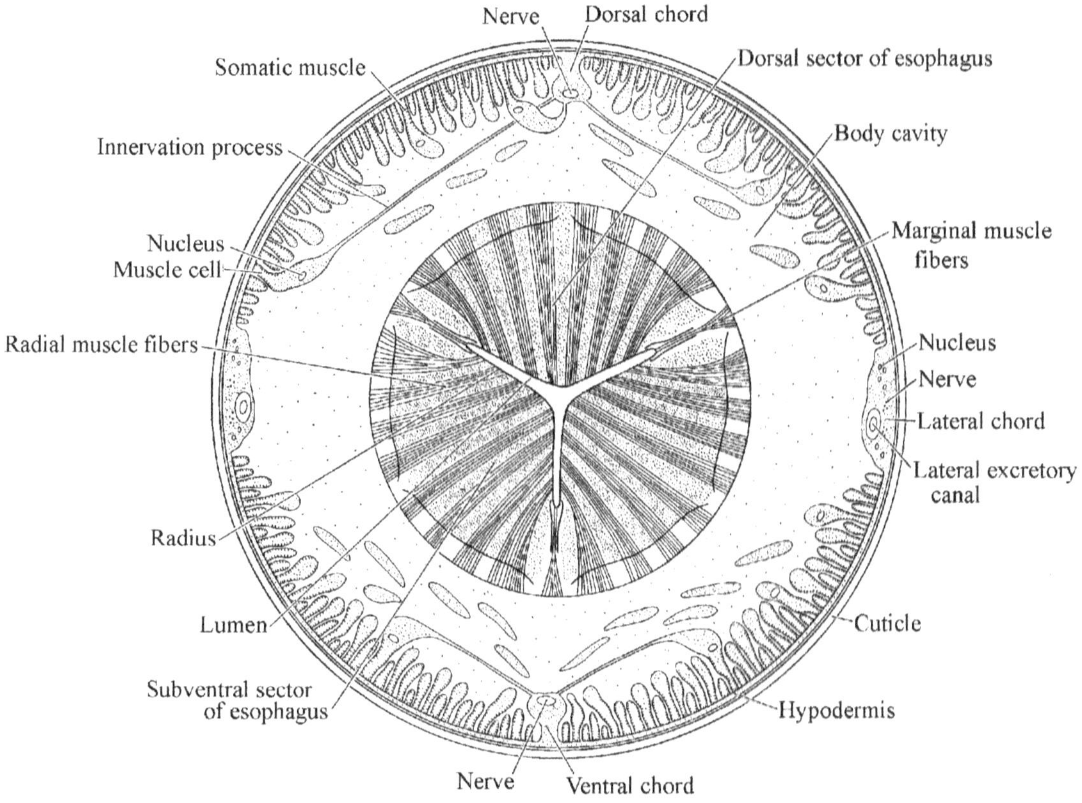

Fig. 6.1 Cross-section of *Ascaris lumbricoides* body wall at the level of the esophagus (after Hirschmann).

3. Why are the subdorsal somatic muscle sectors innervated by the dorsal nerve cord and the subventral somatic muscle sectors innervated by the ventral nerve chord? ______

4. Why do nematodes lie on their side in a dish? ______

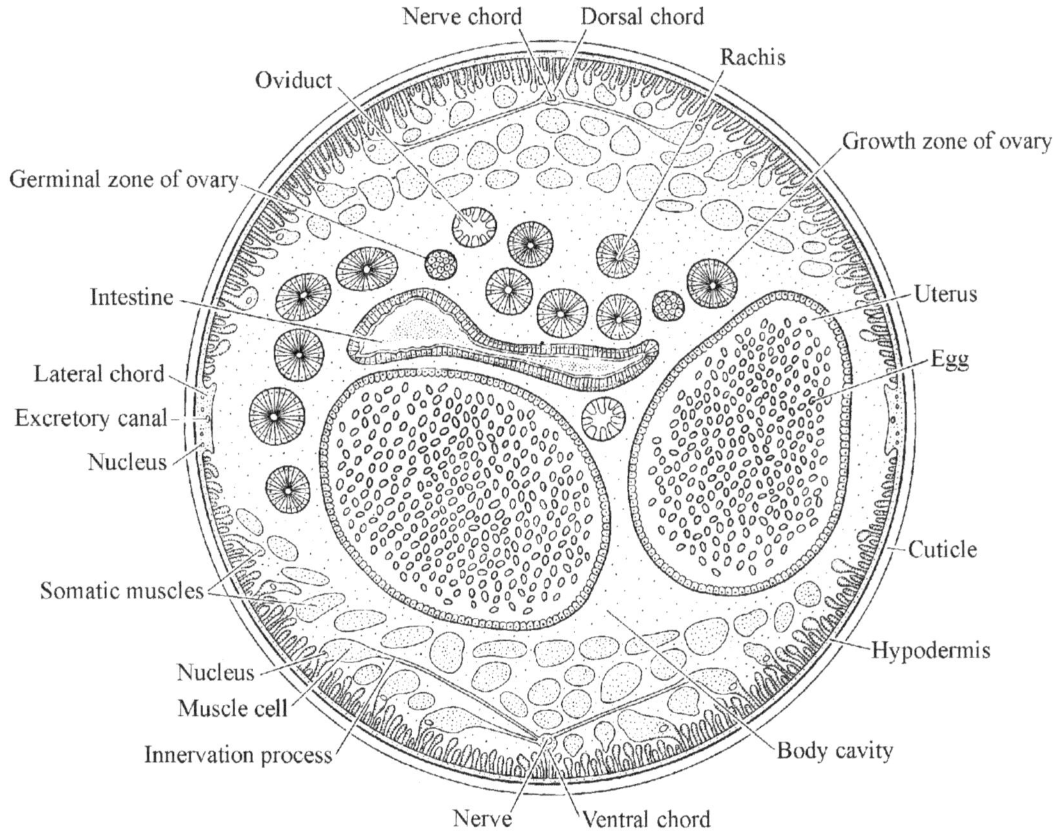

Fig. 6.2 Cross-section of *Ascaris lumbricoides* female body wall (after Hirschmann).

5. *How many hypodermal chords are there in* Ascaris lumbricoides *?* ______________

6. *Why are there so many sections cut through the ovary in the cross section of the body of a female* Ascaris lumbricoides*?* ____

____

____

____

____

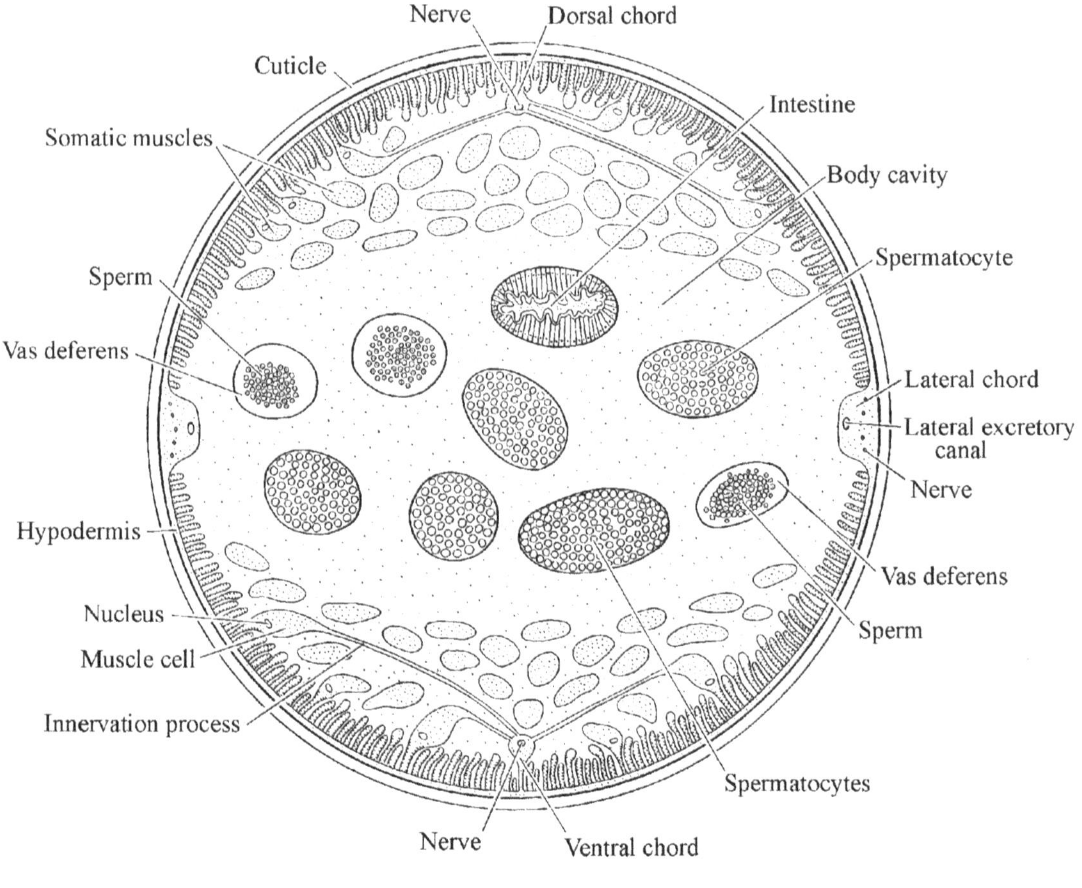

Fig. 6.3 Cross-section of *Ascaris lumbricoides* male body wall (after Hirschmann).

7. Compare and contrast the cross-section of the body wall of the female and male

Ascaris lumbricoides. ____

____

____

____

____

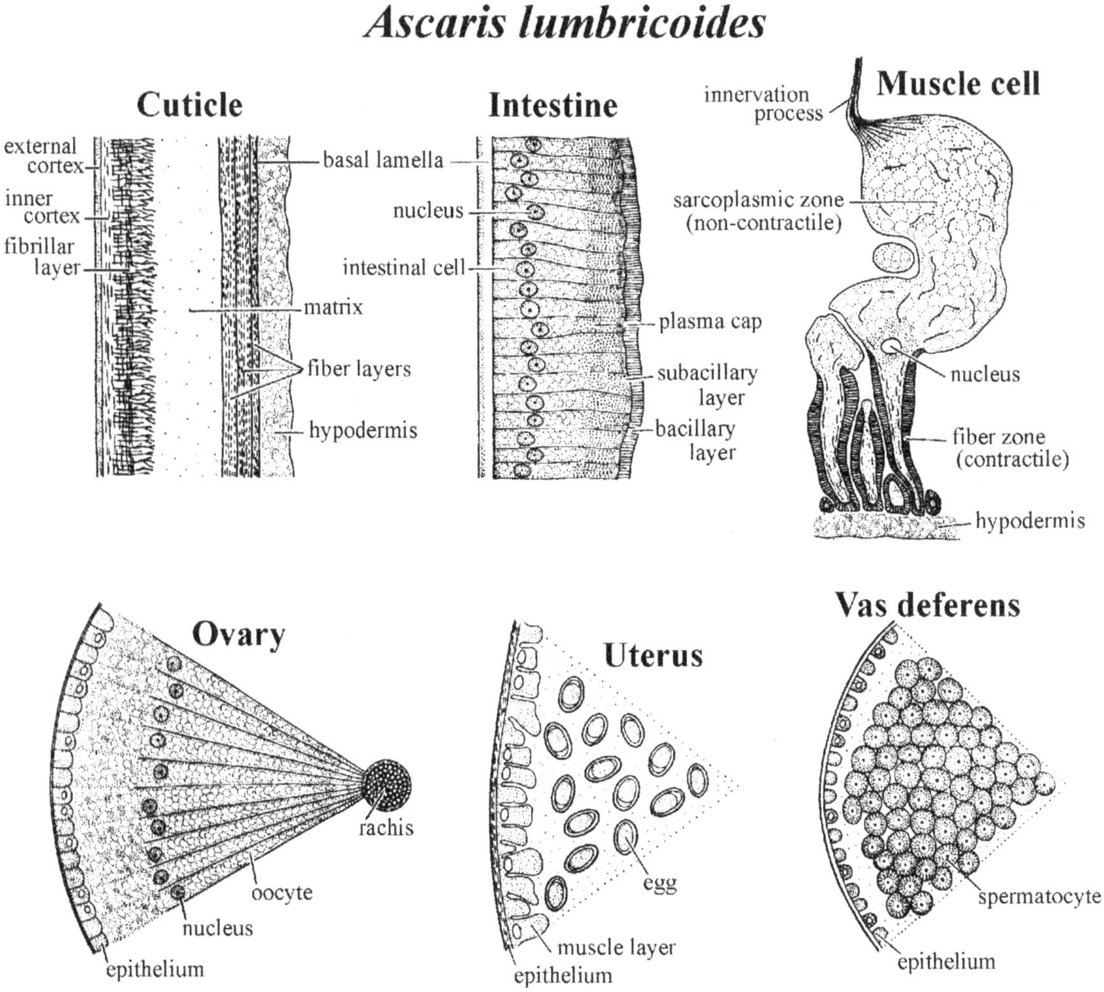

Fig. 6.4 Cross-sections of *Ascaris lumbricoides* body wall, ovary, uterus, and testis (after Hirschmann).

8. *How many sublayers can you discern in the cross-section of* Ascaris lumbricoides?___

____

9. *What is the function of the bacillary layer of the intestine in* Ascaris lumbricoides ? _

____

____

10. What is the purpose of the muscle layer in the uterus of Ascaris lumbricoides *?*_____

Investigation 7

Hypodermis and Somatic Muscles

The hypodermis is a group of cells or a syncytium that secretes the cuticle. In addition, it contains the nerve chords, canals of the excretory/secretory system, and sensory structures. The hypodermis lies between the cuticle and the somatic muscles. It usually forms four longitudinal chords that divide the somatic muscles into four sections. Members of the nematode class Adenophorea are characterized by the occurrence of glands in the hypodermis, whereas they are absent in Secernentea.

Objectives of the Laboratory

Upon completion of this lab the student should be able to:

1. **Observe** the inflated cuticle of *Paratrichodorus* sp.
2. **Draw** the cuticular sheath of *Hemicycliophora* sp.
3. **Sketch** the caudal glands and spinneret of *Mononchus* sp.
4. Illustrate the thick hypodermis of *Heterodera lespedezae* female.
5. **Stain** the ventral nerve chord nuclei in *Rhabditis* sp.
6. **Note** the accumulation of gylcogen in the noncontractile portion of the somatic muscle of *Ascaris* sp.

Materials needed for this laboratory:

Nematode species:

Ascaris sp. (prepared slides)
Hemicycliophora sp.
Heterodera lespedezae
Mononchus sp.
Paratrichodorus sp.
Rhabditis sp.

For each student:

Forceps
Light microscope
Nematode pick
Slide storage box

For each pair of students:

Acid fuchsin lactophenol
Alcohol lamp
Butyl acetate
Cheesecloth
2% formalin
Glass coverslips
Glass microscope slides
Immersion oil
Nail polish
Matches
Proprionic orcein stain
Slide labels
Slide ringer
Small paint brush
Syracuse dishes
Water

Work Sheet

I. Observe the inflated cuticle of *Paratrichodorus* sp. and make an outline sketch of the whole nematode.

II. Note the cuticular sheath of *Hemicycliophora* sp. and draw the tail region where the sheath is distinct.

1. Compare and contrast the inflated cuticle of Paratrichodorus *with the cuticular sheath of* Hemicycliophora. ____

____

____

____

____

____

____

____

III. Draw the caudal glands and the spinneret of *Mononchus* sp.

2. What is the function of the caudal glands and spinneret in Mononchus *sp.?* _______

____

____

____

____

3. If Mononchus *sp. have hypodermal glands, to which class of nematodes do they belong?* ____

IV. Observe the thick hypodermis with numerous nuclei (stained with acid fuchsin-lactophenol) in the fourth-molt female of *Heterodera lespedezae*.

4. Why does the hypodermis become thicker during the molting process? ___________

____

____

____

____

V. Stain males of *Rhabditis* sp. with 1% propionic orcein and observe the nuclei in the ventral hypodermal chord. Note also the muscle nuclei. Specimens should be in lateral position.

A. Procedure: Kill specimens in water by gentle heat. Transfer them to a drop of orcein on a ringed slide. Put coverslip and heat 2-3 seconds over alcohol flame.
B. Remove excess stain.
C. Seal mount with nail polish.

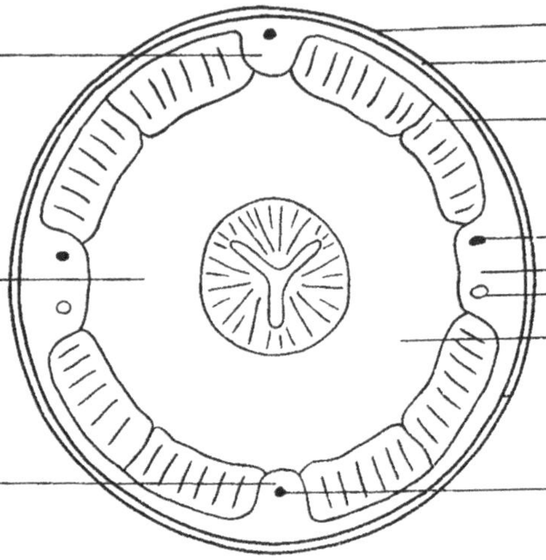

Fig. 8.1 Cross-section of a nematode at the esophageal region.

5. *How many lateral chords do most nematodes have?* ______________

6. *Why is it necessary for the specimen to lie in a lateral position in order to see the ventral hypodermal chord nuclei?* ______________

____

____

7. *What does propionic orcein stain?* ______________

____

VI. Note the accumulation of glycogen (stained purple) in the noncontractile part of the muscles of *Ascaris* sp.

8. *Why does glycogen accumulate in the noncontractile portion of the somatic muscles?*

____

____

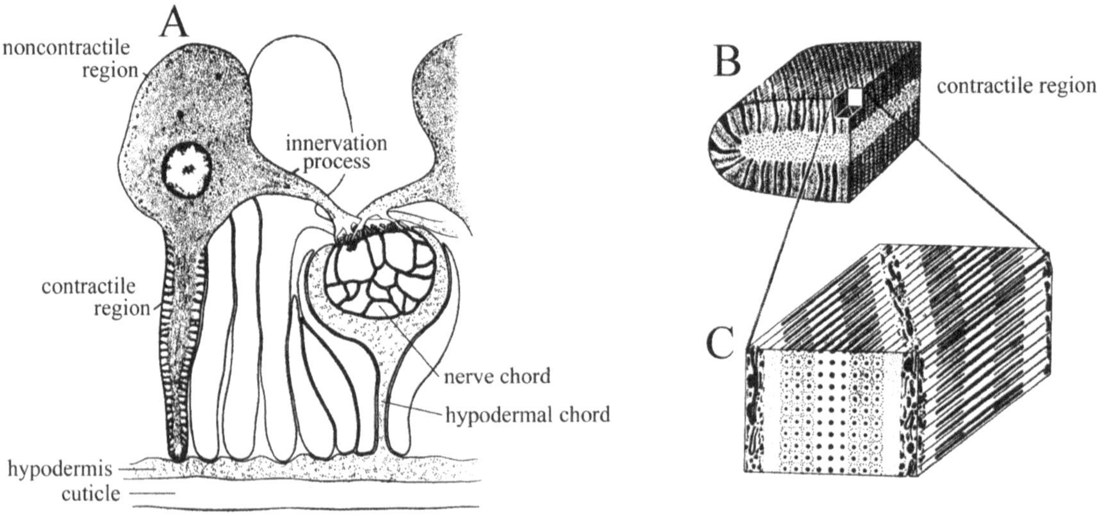

Fig. 8.2 Transverse section of muscle cell of *Ascaris lumbricoides*. A. Section through somatic muscles and hypodermal chord B. Fine structure of contractile region of muscle cell C. Enlargement of one ridge showing the arrangement of the myofilaments (after Rosenbluth)

Investigation 8

Stomas

The stoma or buccal cavity of nematodes is extremely variable in morphology and size, and thus very valuable for systematics and taxonomy. The structure of the stoma provides much information about the feeding habit and mode of parasitism in the case of plant-parasitic nematodes. All plant-parasitic nematodes are armed with a protusible hypodermic needle-like stylet. The stylet enables the nematode to become a plant parasite. Not all nematodes that possess a stylet are plant parasites; however, all plant-parasitic nematodes have a stylet. Nematodes with very long stylets are usually deep feeding ectoparasites; those with short stylets are either endoparasites that migrate into plant tissue in order to feed, or they may be shallow feeding ectoparasites.

Objectives of the Laboratory

Upon completion of this lab the student should be able to:

1. **Observe** stoma in various free-living and plant-parasitic nematode species.
2. **Draw** and label stoma of many different types of nematodes.
3. **Compare** and contrast the stoma, especially the stylet, in Adenophorea with that of Secernentea.

Materials need for this laboratory:

Nematode species:

Belonolaimus sp.
Dorylaimus sp.
Hoplolaimus sp
Mononchus sp.
Panagrolaimus sp
Plectus sp
Pratylenchus sp.
Rhabditis sp.
Trichodorus sp.
Xiphinema sp.

For each student:

- Forceps
- 2% formalin
- Glass coverslips
- Glass microscope slides
- Immersion oil
- Nail polish
- Nematode pick
- Light microscope
- Slide storage box
- Water

Work Sheet

I. Study the different stoma types and make drawings of each:

A. Adenophorea:

1. *Plectus* sp. Represents the Adenophorean version of *Rhabditis*.

2. *Mononchus* sp. Stoma subglobular, heavily sclerotized stomatorhabdions with dorsal tooth.

3. *Dorylaimus* sp. Provided with a protrusible odontostyle which may have originated from a subventral tooth (not hollow-*Nygolaimus*) but through further development became axial and hollow. The odontostyle is formed from a single cell in the wall of the esophagus during postembryogenesis.

4. *Xiphinema* sp. Stylet tremendously elongated and terminating in 3 flanges. *Xiphinema* is a case of total convergence with tylenchids in stylet gross morphology. The flanges and odontostyle extension are differentiations of the esophageal lining. The extent of the odontostyle is indicated by a joint. A guiding ring occurs at the base of odontostyle.

5. *Trichodorus* sp. Protrusible, tooth-like odontostyle, much elongated and curved.

B. Secernentea:

1. *Rhabditis* sp. Stoma shows division into cheilostom, protostom, and telostom.

2. *Panagrolaimus* sp. Stoma moderately wide. Cheilostom slightly sclerotized. Pro- and mesostom distinct and well sclerotized. Metastom with minute denticles. Telostom narrow.

3. *Pratylenchus* sp. The meta- and telo-rhabdions fuse to form a protrusible stylet (stomatostyle). The metarhadbions give rise to the conical part of the stylet, the telorhabdions to stylet shaft and knobs. The cheilo-rhabdions, together with the pro-meso-rhabdions, provide the guiding apparatus. The stomatostyle is formed from myoepithelial cells of the anterior esophagus *in situ* during postembryogenesis.

4. *Hoplolaimus* sp. The stylet is massive with large stylet knobs.

5. *Belonolaimus* sp. The stylet is much elongated.

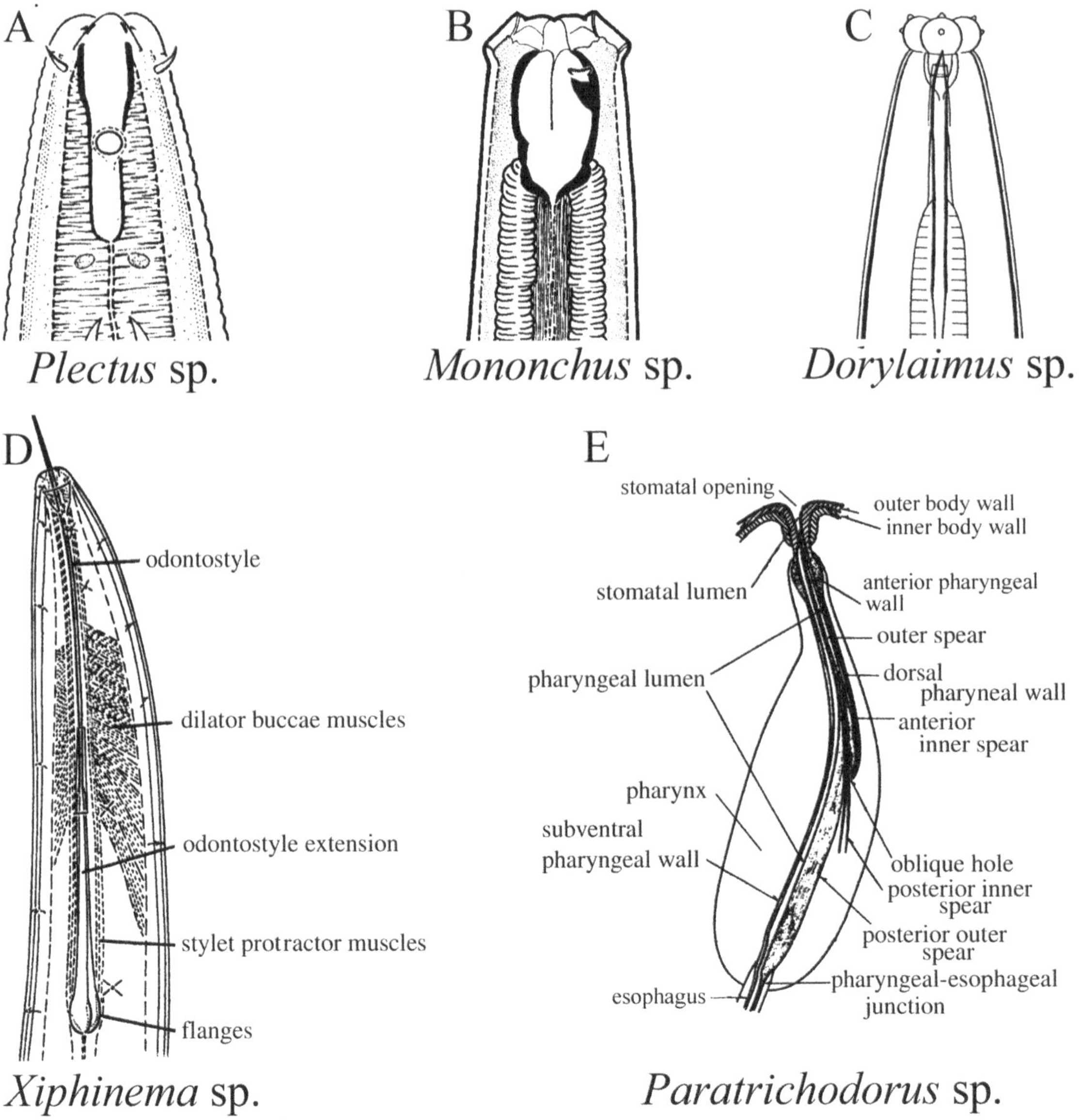

Fig. 8.1. Feeding structures of Adenophorean nematodes. A. *Plectus* sp. B. *Mononchus* sp. (after Jensen and Mulvey) C. *Dorylaimus* sp. (after Hirschmann) D. *Xiphinema* sp. (after Luc) E. *Paratrichordorus minor* (after Hirschmann)

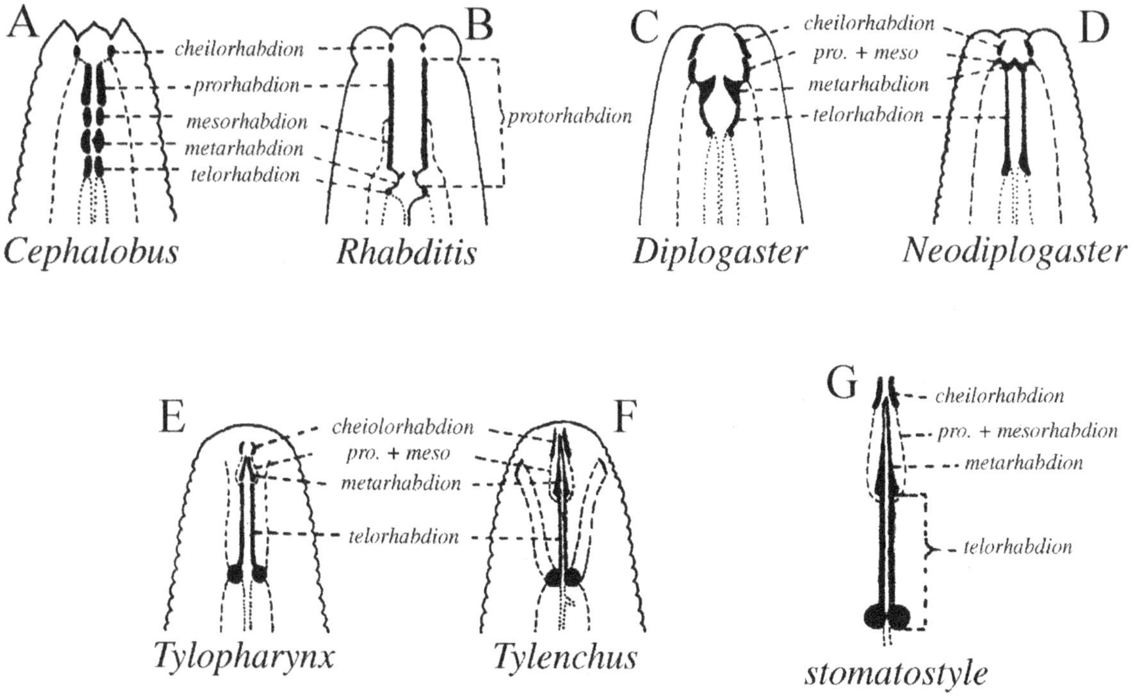

Fig. 8.2. Feeding structures of Secernentean nematodes. A. *Cephalobus* sp. B. *Rhabditis* sp. C. *Diplogaster* sp. D. *Neodiplogaster* sp. E. *Tylopharynx* sp. F. *Tylenchus* sp. G. Morphology of a stomatostyle (after Hirschmann)

1. *What is the purpose of the dorsal tooth in the stoma of* Mononchus *spp.?* __________

____

____

____

2. *Compare and contrast the function of the protusible spear in* Dorylaimus *spp.* ______

____

____

____

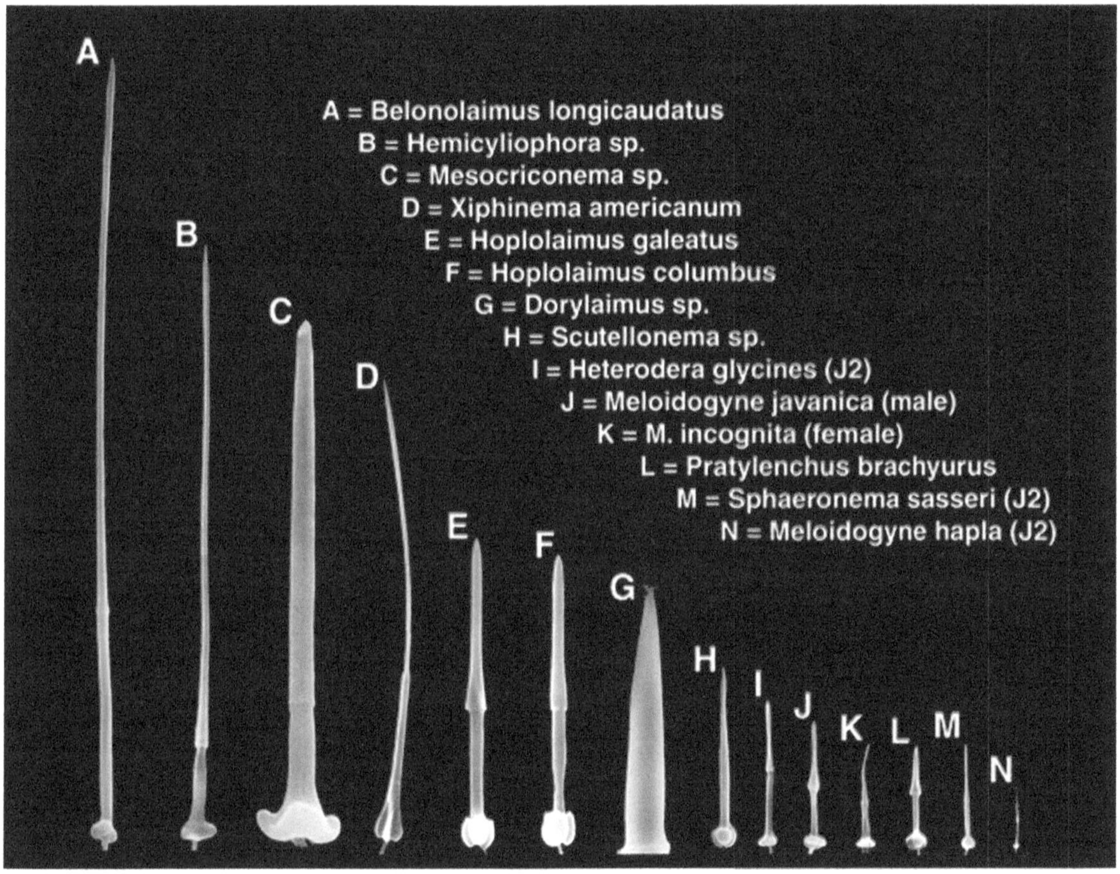

Fig. 8.3 Scanning electron micrographs of excised stylets of various nematodes including mainly plant parasites (after Eisenback and Rammah).

3. Compare and contrast the stylets of Pratylenchus *spp. and* Belonolaimus *spp.* ______

___

___

___

4. Compare and contrast the odontostyle of Xiphinema *spp. with the stomatostyle of*

Belonolaimus *spp.*______

___

___

Investigation 9

Secernentean Stomas

The stoma or buccal cavity of Secernentean nematodes provides valuable information about the feeding habits and modes of parasitism. Some of the free-living and predatory members of the class have cylindrical stomas, globular bucal cavities which may be armed with various kinds of teeth, or a combination of the two forming a globular anterior portion with a cylindrical posterior.

Most of the plant-parasitic nematodes belong to this class and are armed with a unique protusible, hollow, hypodermic needle-like stylet. The stylet enables the nematode to establish a host-parasite relationship with the plant. Nematodes with very long stylets are deep-feeding ectoparasites, and those with short stylets are often endoparasites that migrate into plant tissue in order to feed, or they feed at the surface on root hairs and epidermal cells.

Objectives of the Laboratory

Upon completion of this lab the student should be able to:

1. **Observe** stomas in various free-living and plant-parasitic nematode species.
2. **Draw** and **label** stomas from various Secernentean species.
3. **Compare** and **contrast** stomas from free-living nematodes with that of plant-parasitic nematodes.

Materials needed for this laboratory:

Nematode species:

Acrobeloides sp.
Aphelenchus sp..
Belonolaimus sp
Diplogaster sp.
Ditylenchus sp.
Helicotylenchus sp.
Hoplolaimus sp.
Panagrolaimus sp.
Pratylenchus sp.
Rhabditis sp.

For each pair of students:

- Alcohol lamp
- Butyl acetate
- Cheesecloth
- 2% formalin
- Glass coverslips
- Glass microscope slides
- Immersion oil
- Matches
- Nail polish
- Slide labels
- Slide ringer
- Small paint brush
- Syracuse dishes
- Water

Work Sheet

I. Study the different stoma types in Secernentea and make drawings.

A. *Rhabditis* sp. Stoma shows division into cheilostom, protostom, and telostom.

B. *Acrobeloides* sp. Stoma narrow. The protostom is further subdivided into pro-, meso-, and metastom.

C. *Panagrolaimus* sp. Stoma moderately wide. Cheilostom slightly sclerotized. Pro- and mesostom distinct and well sclerotized. Metastom with minute denticles. Telostom narrow.

D. *Diplogaster* sp. Stoma wide. The metastom bears one or two large teeth, further developmental stages of the minute teeth found on the metastom in *Rhabditis* and *Panagrolaimus*.

E. *Pratylenchus* sp. The meta- and telo-rhabdions fuse to form a protrusible stylet (stomatostyle). The meta-rhadbions give rise to the conical part of the stylet, the telo-rhabdions to stylet shaft and knobs. The cheilo-rhabdions together with the promeso-rhabdions provide the guiding apparatus. The stomatostyle is formed from myo-epithelial cells of the anterior esophagus *in situ* during postembryogenesis.

F. *Aphelenchus* sp. The stylet is delicate without basal knobs.

G. *Ditylenchus* sp. The stylet is short and delicate with three small stylet knobs.

H. *Hoplolaimus* sp. The stylet is massive with large stylet knobs.

I. *Belonolaimus* sp. The stylet is much elongated.

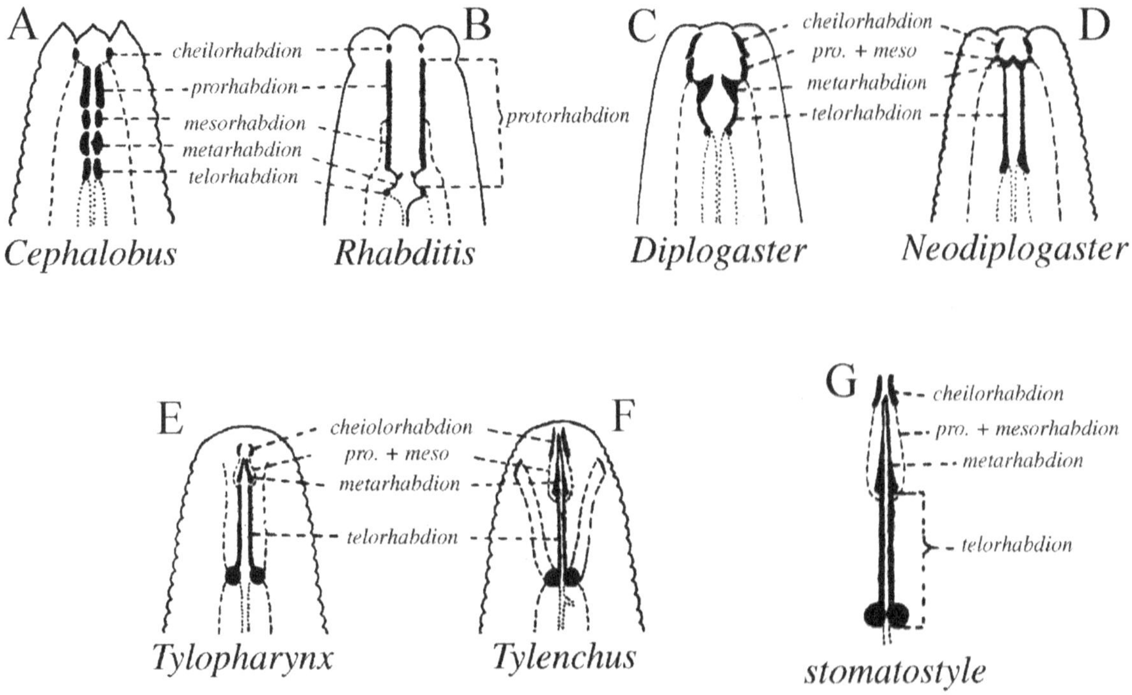

Fig. 9.1 Different stoma types in the nematode class Secernentea (after Hirschmann).

1. Compare and contrast the stomas of Rhabditis *and* Acrobeles. ____________________

____

____

____

2. Compare and contrast the stoma of Rhabditis *and* Diplogaster. ____________________

____

____

____

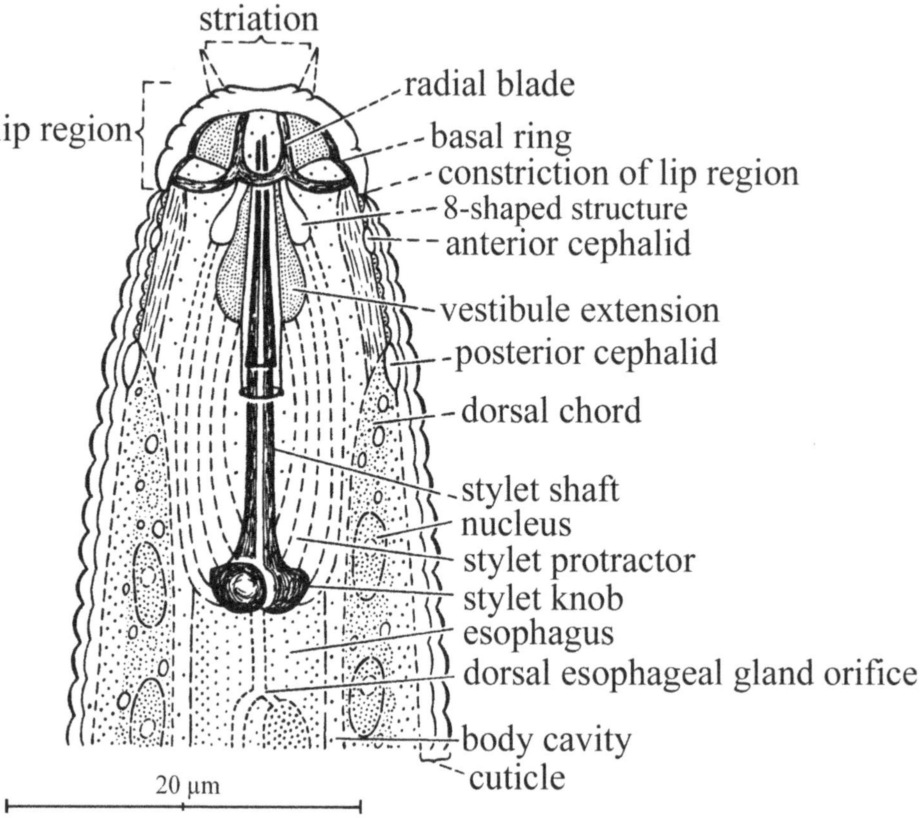

Fig. 9.2 Cephalic region of a male of soybean cyst nematode, *Heterodera glycines* (after Hirschmann).

3. *Compare and contrast the style of* Aphelenchus spp. *with that of* Ditylenchus spp. ___

---

4. *Compare and contrast the stylet of* Hoplolaimus spp. *with that of* Belonolaimus spp. _

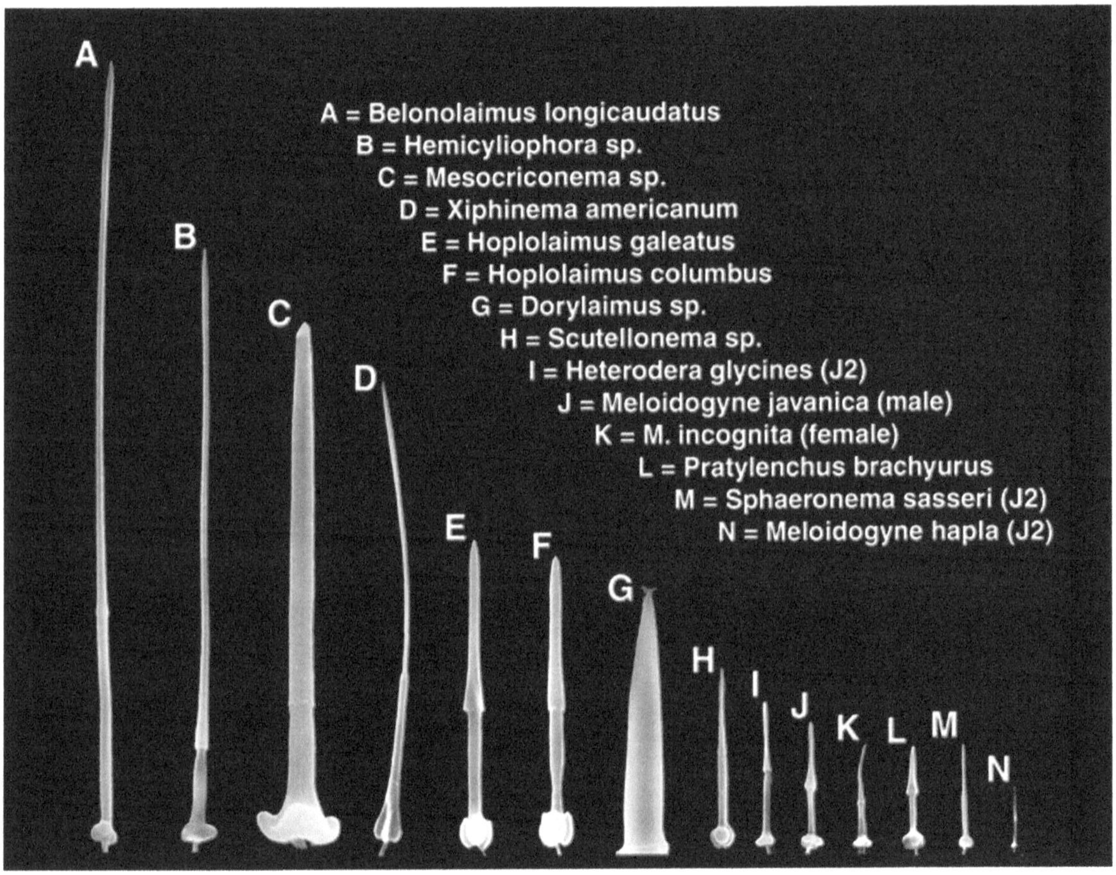

Fig. 9.3 Scanning electron micrographs of excised stylets of various genera, species, and life-stages of nematodes.

5. *Based on morphology of the stoma, what type of feeding behavior does* Acrobeloides *spp. have?* ____________________

6. *Based on stylet length, is* Anguina *a deep feeder or a migratory endoparasite?* ____________________

____

7. *What is the function of the radial blades and basal ring of the cephalic framework of a stylet-bearing nematode?* ____________________

8. *Where does the stylet opening occur in the Secernentean nematodes?* ____________________

____

Investigation 10

Adenophorean Stomas

The stoma or buccal cavity of Adenophorean nematodes is extremely variable in morphology and size, and thus very valuable for systematics and taxonomy. The structure of the stoma provides information about the feeding habit and mode of parasitism. Most of the members of this class are free-living, predators, or animal parasites. A few of them are plant parasites which are armed with a protusible hypodermic needle-like stylet. Nematodes with very long stylets are deep-feeding ectoparasites, and those with short stylets are shallow-feeding ectoparasites.

Objectives of the Laboratory

Upon completion of this lab the student should be able to:

1. **Observe** stomas in various free-living and plant-parasitic nematode species.
2. **Draw** and **label** stomas from various Adenophorean species.
3. **Compare** and **contrast** stomas from free-living nematodes with that of plant-parasitic nematodes.

Materials needed for this laboratory:

Nematode species:

Dorylaimus sp.
Mononchus sp.
Plectus sp.
Longidorus sp.
Paratrichodorus sp.
Xiphinema sp.

For each student:

Forceps	Nematode pick
Light microscope	Slide storage box

For each pair of students:

Alcohol lamp	Matches
Butyl acetate	Nail polish
Cheesecloth	Slide labels
2% formalin	Slide ringer
Glass coverslips	Small paint brush
Glass microscope slides	Syracuse dishes
Immersion oil	Water

Work Sheet

I. Study the following types of stoma types of Adenophorea and make drawings:

A. *Plectus* sp. Represents the adenophorean version of *Rhabditis.*

B. *Mononchus* sp. Stoma subglobular, heavily sclerotized stoma-rhabdions with a dorsal tooth.

C. *Dorylaimus* sp. Provided with a protrusible odontostyle which may have evolved from a subventral tooth (not hollow in *Nygolaimus* sp.) but through further development became axial and hollow. The odontostyle is formed from a single cell in the wall of the esophagus during post embryogenesis.

D. *Xiphinema* sp. Stylet tremendously elongated and terminating in 3 flanges. *Xiphinema* is a case of total convergences with tylenchids in stylet gross morphology. The flanges and ondontosyle extension are differentiations of the esophageal lining. The extent of the odontostyle is indicated by a joint. The guiding ring occurs at the base of the odontostyle.

E. *Longidorus* sp. Odontostyle greatly elongated. Odontostyle extension up to 2/3 the length of the odontostyle with lumen widening slightly near the base. No flanges. The guiding ring is located near the anterior end of the odontostyle.

F. *Trichodorus* or *Paratrichodorus* sp. Protusible, tooth-like odontostyle, much elongated and curved.

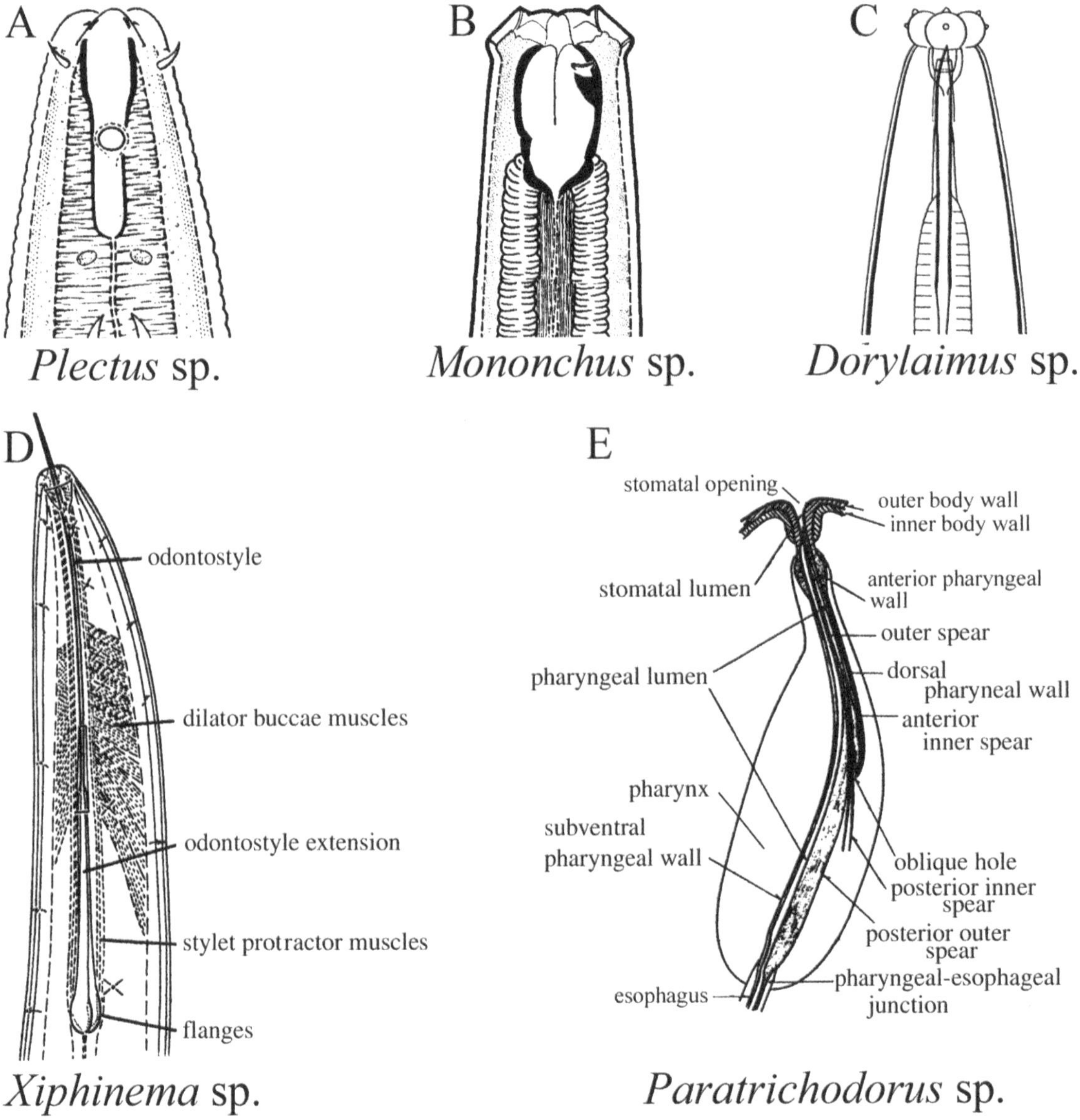

Fig. 10.1. Feeding structures of Adenophorean nematodes. A. *Plectus* sp. B. *Mononchus* sp. (after Jensen and Mulvey) C. *Dorylaimus* sp. (after Hirschmann) D. *Xiphinema* sp. E. *Paratrichodorus* sp

1. *Compare and contrast the stoma of* Plectus *spp. and* Mononchus *spp.* ______________

____

____

2. *Compare and contrast the stylet of* Dorylaimus *spp. and* Xiphinema. *spp.* __________

3. *Compare and contrast the stylet of* Xipinema *spp. with that of* Longidorus *spp.*______

4. *Based on stylet mophology, compare and contrast the mode of parasitism of* Xiphinema spp. *and* Longidorus *spp.* __________

5. *Based on stylet morphology, what is the mode of parasitism of* Paratrichodorus *spp*?

Investigation 11

Esophagi

The esophagi of nematodes are diverse structures and useful for identification. Basically a pumping organ, the esophagus is composed of muscles and gland cells that surround a triradiate lumen lining with 3-5 interspersed esophageal gland cells. A one-part esophagus consists of a muscular cylinder with several gland cells contained within. A two-part esophagus consists of a narrow muscular anterior portion and a swollen, glandular basal portion. A four-part esophagus contains an anterior corpus, a swollen metacorpus or median bulb, a narrow isthmus, and an enlarged glandular posterior bulb. In a three-part esophagus the procorpus and metacorpus are fused together. Specialized structures for grinding food particles may be found in the basal bulb. The basal gland lobe may join with the intestine in a line perpendicular to the body or it may overlap it dorsally or ventrally. An esophageal-intestinal valve connects the esophagus with the intestine. It is triangular and is often surrounded by a circular sphincter muscle which prevents food flowing down the esophagus into the intestine from backing up.

Objectives of the Laboratory

Upon completion of this lab the student should be able to:

1. **Identify** the major types of esophagi in the nematode classes Adenophorea and Secernentea.
2. **Draw** and **label** the major types of esophagi in the nematode classes Adenophorea and Secernentea.

Materials needed for this laboratory:

Nematode species:

Aphenlenchoides sp.
Dorylaimus sp.
Mononchus sp.
Rhabditis sp.
Diplogaster sp.
Mesocriconema sp.
Plectus sp.
Tylenchorhynchus sp.

For each student:

Forceps
Light microscope
Nematode pick
Slide storage box

For each pair of students:

Alcohol lamp
Baermann funnels
Butyl acetate
Cheesecloth
2% formalin
Glass coverslips
Glass microscope slides
Immersion oil
Matches
Nail polish
Slide labels
Slide ringer
Small paint brush
Syracuse dishes
Water

Work Sheet

I. Study the following types of esophagi of Adenophorea and make drawings.

A. *Mononchus* sp. Cylindrical type. Five uninucleate esophageal glands: one dorsal, 4 subventral. Esophago-intestinal valve triradiate and massive.

B. *Dorylaimus* sp. Dorylaimoid type. Esophagus consisting of cylindrical corpus followed by an elongate cylindrical glandular region. Five uninucleate esophageal glands: one dorsal, 4 subventral. Esophago-intestinal valve well developed.

C. *Plectus* sp. Plectoid type. Esophagus consisting of cylindrical corpus and pyriform bulbar region containing a triradiate, denticulate valve apparatus. Three uninucleate esophageal glands: 1 dorsal, 2 subventral. Esophago-intestinal valve elongate, conoid.

II. Study the following types of Secerenentean esophagi and make drawings.

A. *Rhabditis* sp. Rhabditoid type. Esophagus consisting of corpus (which is divided into cylindrical procorpus and enlarged, distinctly set-off metacorpus),

B. *Diplogaster* sp. Diplogasteroid type. Esophagus consisting of corpus (which is divisible into cylindrical corpus and enlarged distinctly set-off metacorpus), narrow isthmus and pyriform bulbar region without a valve. Three uninucleate esophageal glands: one dorsal, two subventral. Esophago-intestinal valve present.

C. *Tylenchorhynchus* sp. Tylenchoid type. Esophagus consisting of cylindrical procorpus, enlarged metacorpus and bulbar region. Muscles of this region degenerated, esophageal glands (3) from the entire bulbar region. Esophago-intestinal valve (cardia) present. Dorsal gland orifice situated in the anterior part of the procorpus.

D. *Mesocriconema* sp. Criconematoid type. Esophagus consisting of a fused procorpus and metacorpus, isthmus, gland lobe region with 3 glands extending back over the anterior end of the intestine. Dorsal gland orifice situated in the anterior part of the procorpus.

E. *Aphelenchoides* sp. Aphelenchoid type. Esophagus consisting of cylindrical procorpus, enlarged metacorpus and esophageal glands (3) extending back over the anterior end of the intestine in a single lobe. Dorsal gland orifice situated in the anterior part of the metacorpus.

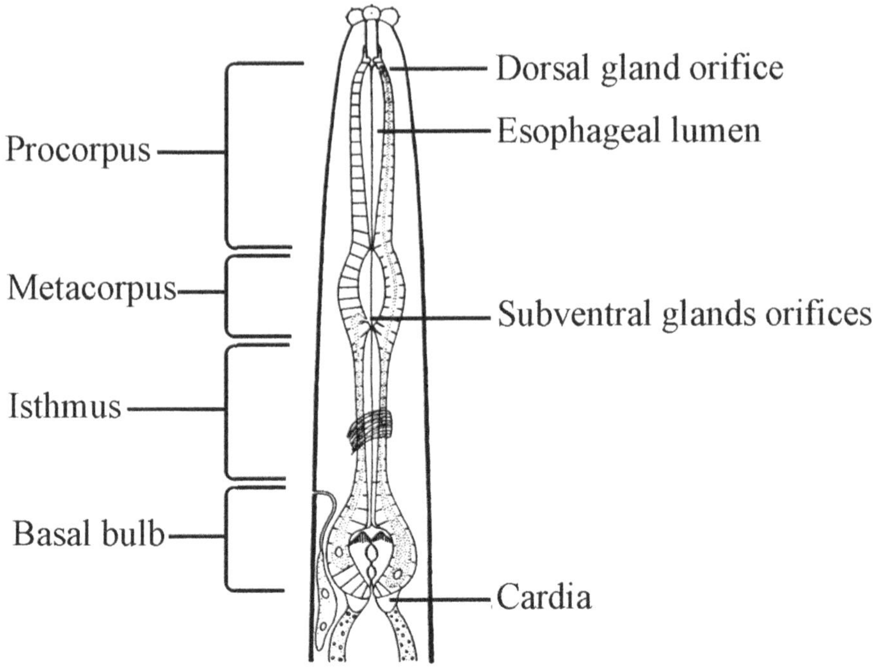

Fig. 11.1 Morphology of a nematode esophagus (after Hirschmann).

1. What are the basic types of nematode esophagi? ______________________________

____

____

2. How is esophageal morphology related to feeding habits? ______________________

____

____

____

____

____

3. Where are the glands located in a cylindrical type of esophagus? ________________

____

____

*4. What is the purpose of the denticulate pumping apparatus in a bulboid esophagus?*___

____

____

5. Compare and contrast a cylindrical and a dorylaimoid type of esophagi. _________

____

____

____

____

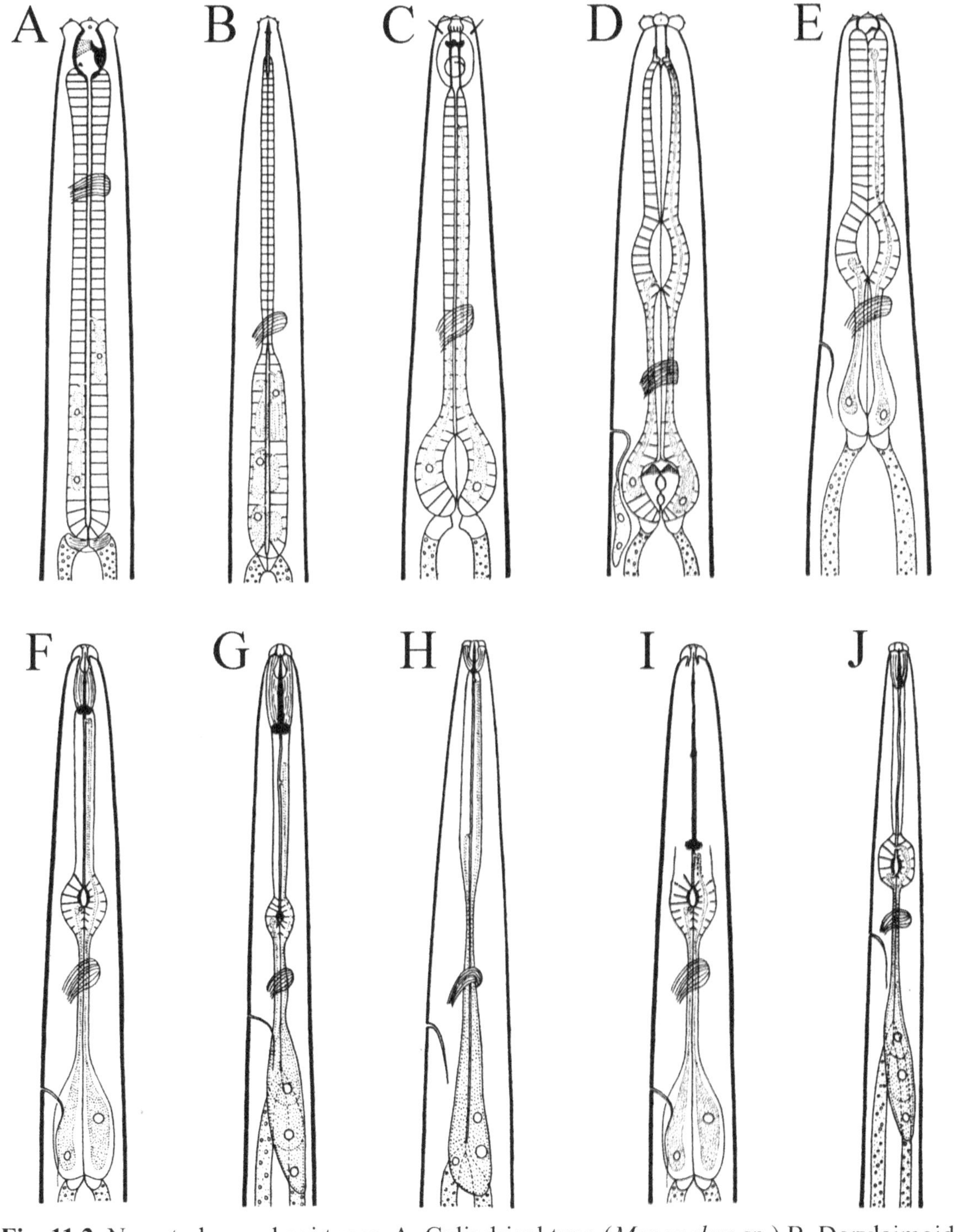

Fig. 11.2 Nematode esophagi types. A. Cylindrical type (*Mononchus* sp.) B. Dorylaimoid type (*Dorylaimus* sp.) C. Bulboid type (*Plectus* sp.) D. Rhabditoid type (*Rhabditis* sp.) E. Diplogasteroid type (*Diplogaster* sp.) F. Tylenchoid type (*Tylenchorhynchus* sp.) G. Tylenchoid type (*Helicotylenchus* sp.) H. Tylenchoid type (*Neotylenchus* sp.) I. Criconematoid type (*Mesocriconema* sp.) J. Aphelenchoid type (*Aphelenchus* sp.) (A-J after Hirschmann)

6. *Compare and contrast a dorylaimoid and plectoid type of esophagi.* ____________

7. *Compare and contrast a plectoid and rhabditoid type of esophagi.* ____________

8. *Compare and contrast a rhabditoid and diplogasteroid type of esophagi.* ____________

9. *Compare and contrast a criconematoid and tylenchoid type of esophagi.* ____________

10. *Compare and contrast the various types of tylenchoid esophagi.* ____________

Investigation 12

Secernentean Esophagi

Nematodes within the class Secernentea have various types of esophagi. The major types include those with one-part that consists of a muscular cylinder with several gland cells contained within; a two-part esophagus that consists of a narrow muscular anterior portion and a swollen, glandular basal portion; and a four-part esophagus that contains an anterior corpus, a swollen metacorpus or median bulb, a narrow isthmus, and an enlarged glandular posterior bulb. In a three-part esophagus, the procorpus and metacorpus are fused together. Specialized structures for grinding food particles may be found in the basal bulb. The basal gland lobe may join with the intestine in a line perpendicular to the body or it may overlap it dorsally or ventrally. An esophageal-intestinal valve connects the esophagus with the intestine. It is triangular and is often surrounded by a circular sphincter muscle which prevents food flowing down the esophagus into the intestine from backing up.

Objectives of the Laboratory

Upon completion of this lab the student should be able to:

1. **Identify** the major types of esophagi in the class Secernentea.
2. **Draw** and **label** the major types of esophagi in the class Secernentea.

Materials needed for this laboratory:

Nematode species:

Aphelenchus sp
Helicotylenchus sp.
Panagrellus sp.
Tylenchorhynchus sp.
Diplogaster sp.
Hoplolaimus sp.
Rhabditis sp
Mesocriconema sp.

For each student:

Forceps
Light microscope
Nematode pick
Slide storage box

For each pair of students:

Alcohol lamp
Butyl acetate
Cheesecloth
2% formalin
Glass coverslips
Glass microscope slides
Immersion oil
Matches
Nail polish
Slide labels
Slide ringer
Small paint brush
Syracuse dishes
Water

Work Sheet

I. Study the different types of esophagi in the following Secernentea and make drawings:

A. *Panagrellus* sp. Rhabditoid type. Esophagus consisting of cylindrical corpus, narrow isthmus and pyriform bulbar region containing a well-developed value. Three uninucleate esophageal glands: 1 dorsal, 2 subventral. Esophago-intestinal valve well developed.

B. *Rhabditis* sp. Rhabditoid type. Esophagus consisting of corpus (which is subdivided into cylindrical procorpus and enlarged distinctly set off metacorpus), narrow isthmus and pyriform bulbar region containing a well-developed valve. Three uninucleate esophageal glands: one dorsal, 2 subventral. Esophago-intestinal valve pronounced.

C. *Diplogaster* sp. Diplogasteroid type. Esophagus consisting of corpus (which is subdivided into cylindrical procorpus and enlarged distinctly set off, valved metacorpus), narrow isthmus and pyriform, glandular bulbar region without valve. Three uninucleate esophageal glands: one dorsal, 2 subventral. Esophago-intestinal valve present.

D. *Tylenchorhynchus* sp. Tylenchoid type. Esophagus consisting of cylindrical procorpus, enlarged, valved metacorpus, slender isthmus and bulbar region. Muscles of latter region degenerated, 3 esophageal glands form the entire bulbar region. Esophago-intestinal valve present. Dorsal gland orifice situated in anterior part of procorpus.

E. *Helicotylenchus* sp. Tylenchoid type. Esophagus consisting of cylindrical procorpus, enlarged, valved metacorpus, slender isthmus, and esophageal glands (3) extending back over anterior end of intestine in 3 lobes (ventral, lateral and dorsal). Dorsal gland orifice situated in anterior part of procorpus.

F. *Hoplolaimus* sp.Tylenchoid type. Esophagus consisting of cylindrical procorpus enlarged, valved metacorpus, slender isthmus, and esophageal glands (3) overlapping intestine dorsally and laterally. Dorsal gland orifice situated in anterior part of procorpus.

G. *Mesocriconema* sp. Criconematoid type. Esophagus consisting of procorpus fused with the metacorpus, slender isthmus, and esophageal glands (3) overlapping esophagus ventrally. Dorsal gland orifice situated in anterior part of procorpus.

H. *Aphelenchus* sp. Aphelenchoid type. Esophagus consisting of cylindrical procorpus, enlarged, valved metacorpus and esophageal glands (3) extending backover anterior end of intestine in a single lobe. Dorsal gland orifice situated in anterior part of metacorpus.

1. Compare and contrast the rhabditoid and diplogasteroid types of esophagi. ________

____

____

____

____

2. Compare and contrast the diplogasteroid and tylenchoid types of esophagi. ________

____

____

____

____

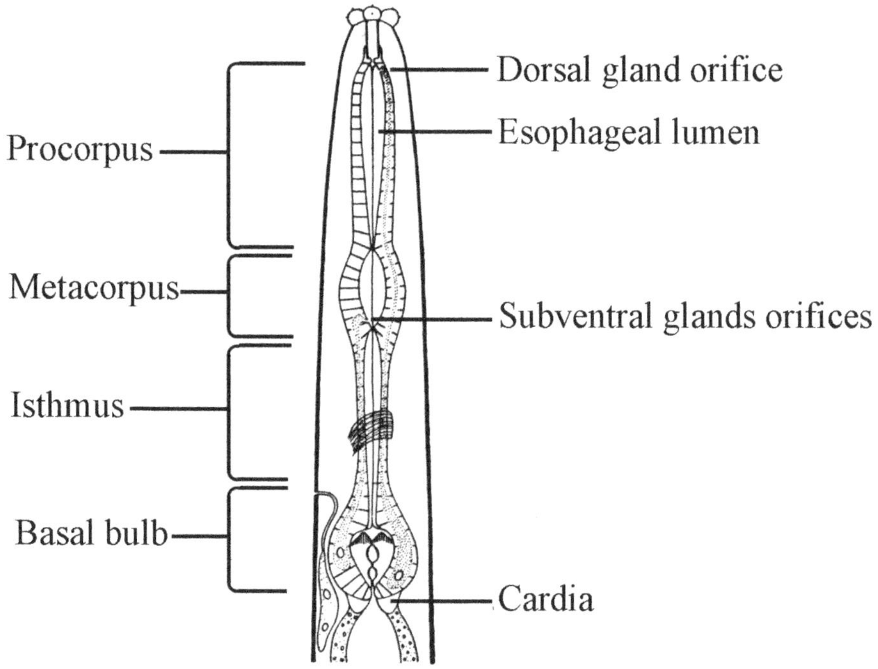

Fig. 12.1 Morphology of a nematode esophagus (after Hirschmann).

3. Compare and contrast the tylenchoid and aphelenchoid types of esophagi. _________

4. Compare and contrast the the tylenchoid and criconematoid types of esophagi. _____

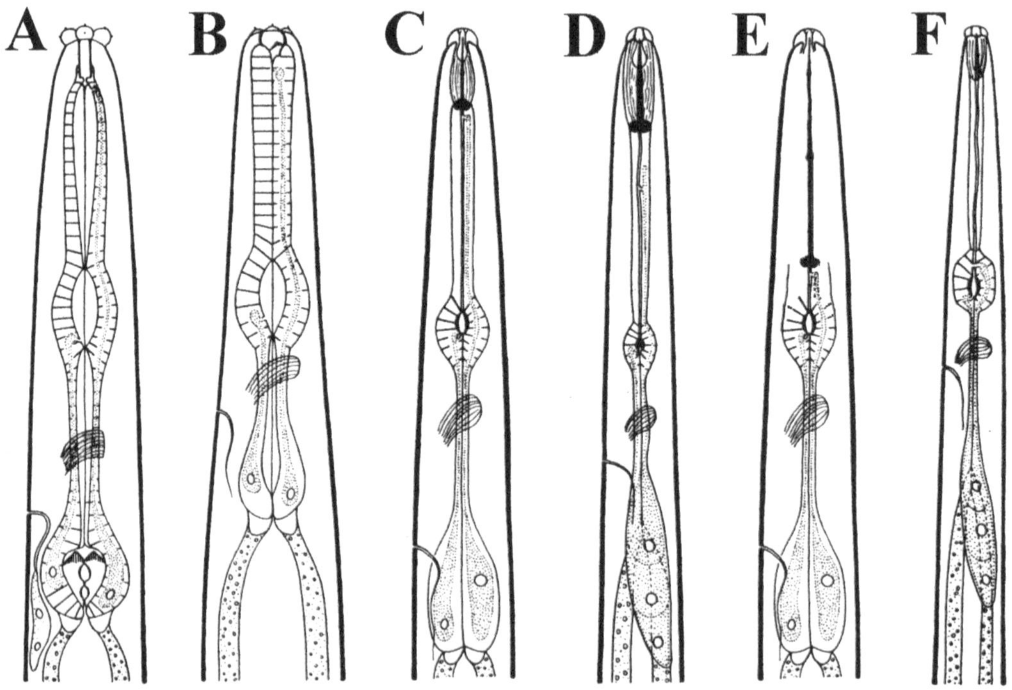

Fig. 12.2 Nematode esophagi types. A. Rhabditoid type (*Rhabditis* sp.) B. Diplogasteroid type (*Diplogaster* sp.) C. Tylenchoid type (*Tylenchorhynchus* sp.) D. Tylenchoid type (*Helicotylenchus* sp.) E. Criconematoid type (*Mesocriconema* sp.) F. Aphelenchoid type (*Aphelenchus* sp.) (A-F after Hirschmann)

5. *Describe some of the variations within the tylenchoid type of esophagi.* ____________

____

____

____

____

6. *What is the purpose of the denticulate basal pump in the rhabditoid type of esophagus?* ____

____

____

Investigation 13

Adenophorean Esophagi

The esophagi of nematodes are diverse structures and useful for identification. Basically a pumping organ, the esophagus is composed of muscles and gland cells that surround a triradiate lumen lining with 3-5 interspersed esophageal gland cells. A one-part esophagus consists of a muscular cylinder with several gland cells contained within. A two-part esophagus consists of a narrow muscular anterior portion and a swollen, glandular basal portion. A four-part esophagus contains an anterior corpus, a swollen metacorpus or median bulb, a narrow isthmus, and an enlarged glandular posterior bulb. In a three-part esophagus, the procorpus and metacorpus are fused together. Specialized structures for grinding food particles may be found in the basal bulb. The basal gland lobe may join with the intestine in a line perpendicular to the body, or it may overlap it dorsally or ventrally. An esophageal-intestinal valve connects the esophagus with the intestine. It is triangular and is often surrounded by a circular sphincter muscle which prevents food, flowing down the esophagus into the intestine, from backing up.

Objectives of the Laboratory

Upon completion of this lab the student should be able to:

1. **Identify** the major types of esophagi in the class Adenophorea.
2. **Draw** and label the major types of esophagi in the class Adenophorea.

Materials needed for this laboratory:

Nematode species:

Dorylaimus sp.
Mononchus sp.
Paratrichodorus sp.
Plectus sp.
Prodesmodora sp.
Xiphinema sp.

For each student:

- Forceps
- Light microscope
- Nematode pick
- Slide storage box

For each pair of students:

- Alcohol lamp
- Baermann funnels
- Butyl acetate
- Cheesecloth
- 2% formalin
- Glass coverslips
- Glass microscope slides
- Immersion oil
- Matches
- Nail polish
- Slide labels
- Slide ringer
- Small paint brush
- Syracuse dishes
- Water

Work Sheet

I. Study the following types of esophagi of Adenophorea and make drawings:

A. *Mononchus* sp. Cylindrical type. Five uninucleate esophageal glands: 1 dorsal, 4 subventral. Esophago-intestinal valve triradiate and massive.

B. *Dorylaimus* sp. Dorylaimoid type. Esophagus consisting of cylindrical corpus followed by an elongate cylindrical glandular region. Five uninucleate esophageal glands: 1 dorsal, 4 subventral. Esophago-intestinal valve well developed.

C. *Xiphinema* sp. Dorylaimoid type. Three uninucleate esophageal glands: 1 dorsal, 2 subventral. Esophago-intestinal valve conoid.

D. *Paratrichodorus* sp. Dorylaimoid type. Esophagus anteriorly cylindrical and slender, expanding gradually posteriorly to form an elongate conoid swelling. Five uninucleate esophageal glands. Esophago-intestinal valve small, conoid.

E. *Prodesmodora* sp. Bulboid type. Esophagus divisible into cylindrical corpus and pyriform bulbar region containing 3 cuticularized crescentic valve elements. Three uninucleate esophageal glands: one dorsal, 2 subventral. Esophago-intestinal valve short and flattened.

F. *Plectus* sp. Plectoid type. Esophagus consisting of cylindrical corpus and pyriform bulbar region containing a triradiate, denticulate valve apparatus. Three uninucleate esophageal glands: 1 dorsal, 2 subventral. Esophago-intestinal valve elongate, conoid.

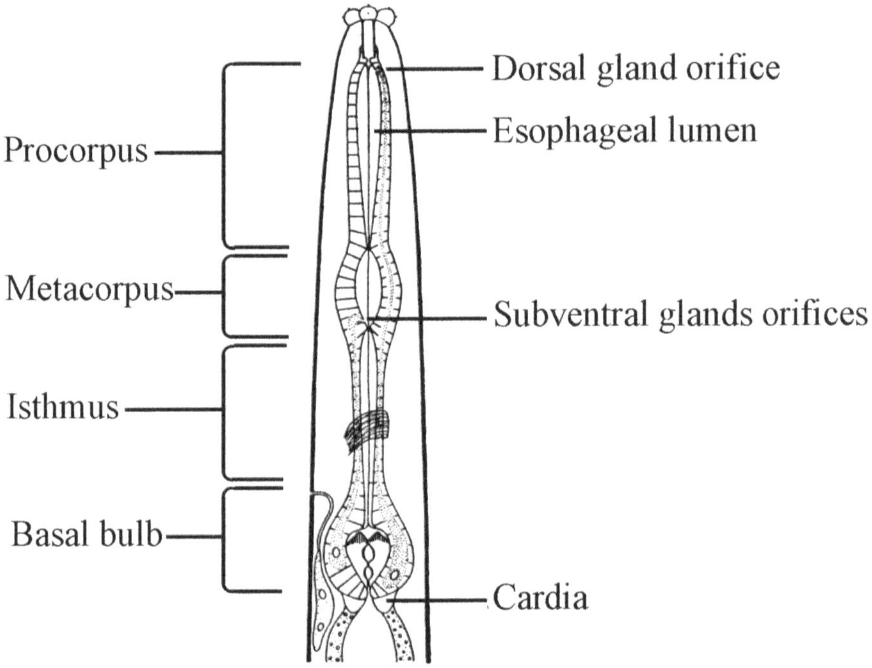

Fig. 13.1 Morphology of a nematode esophagus (after Hirschmann).

1. *Compare and contrast the mononchoid and dorylaimoid types of esophagi.* __________

2. *Compare and contrast the dorylaimoid and bulboid types of esophagi.* __________

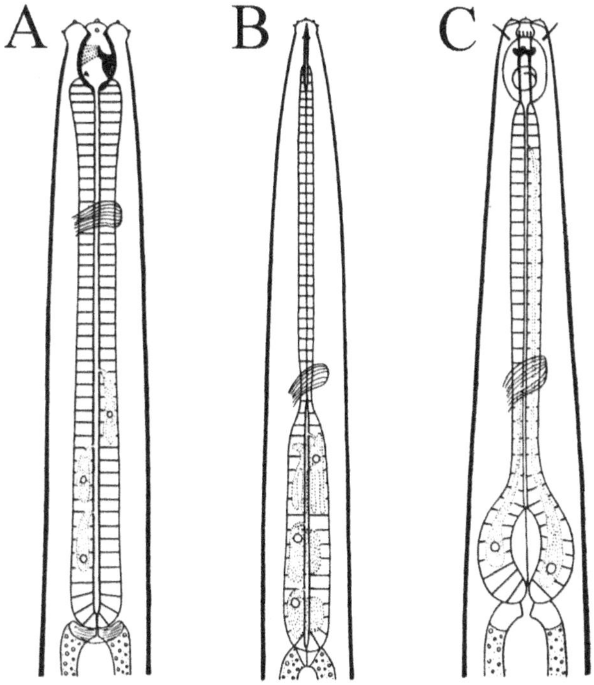

Fig. 13.2 Nematode esophagi types. A. Cylindrical type (*Mononchus* sp.) B. Dorylaimoid type (*Dorylaimus* sp.) C. Bulboid type (*Plectus* sp.)

3. *What is the purpose of the tooth in the mononchoid type of esophagus?* ____________

____

____

____

4. *Where are the esophageal gland cells in each type esophagus examined?* ___________

____

____

____

____

Investigation 14

Intestine

The gross morphology of the intestine of various groups of nematodes is remarkably similar. Generally, the intestine is a simple tube that is formed from a single layer of epithelial cells and fills the body space between the esophagus and the posterior gut. Several plant-parasitic nematodes may have an anterior caecum that extends beyond the base of the esophagus, almost to the level of the nerve ring. A posterior intestinal caecum is extremely rare.

The single layer of epithelial cells often have an inner bacillary layer that greatly increases the surface area of the lumen. This layer varies from having large, well-defined microvilli to ones that are fine and compact. The total number of cells and their shapes are responsible for the overall appearance of the intestine and its lumen. Oligocytous intestines contain up to 128 longitudinally elongate to rectangular cells. Intestines that have 20-50 hexagonal cells in circumference are polycytous; they are myriocytous if they are more than 100 cuboidal cells in circumference. Many plant-parasitic nematodes have an intestine which is a synticium where cell walls are lacking, the lumen is reduced in form or absent, and the microvilli are less compact or absent.

Objectives of the Laboratory

Upon completion of this lab the student should be able to:

1. **Identify** various types of intestines that occur in nematodes.
2. **Draw** and label the major types of inclusions in the intestines of nematodes.
3. **Observe** the bacillary layer and lumen in the nematode intestine.

Materials needed for this laboratory:

Nematode species:

Belonolaimus sp.
Meloidogyne sp. male
Rhabditis sp.
Hoplolaimus sp.
Mononchus sp.

For each student:

Forceps
Light microscope
Nematode pick
Slide storage box

For each pair of students:

Alcohol lamp
Baermann funnels
Butyl acetate
Cheesecloth
2% formalin
Glass coverslips
Glass microscope slides
Immersion oil
Matches
Nail polish
Slide labels
Slide ringer
Small paint brush
Syracuse dishes
Water

Work Sheet

I. Study the intestine of *Rhabditis* sp., *Mononchus* sp., *Hoplolaimus* sp., *Belonolaimus* sp., and *Meloidogyne* males.

A. Note the presence of distinct cells in *Rhabditis* sp. (oligocytous intestine) and *Mononchus* sp. (polycytous intestine), whereas cells are absent in *Hoplolaimus* sp. and *Belonolaimus* sp.

B. Draw and label the sphaerocrystals of rhabditin in the intestinal cells of *Rhabditis* sp. and the fat globules in the intestine of *Hoplolaimus* sp.

C. Note the presence of "intestinal fasciculi" in *Belonolaimus* sp.

D. Observe the anterior caecum in the male root-knot nematode, *Meloidogyne* sp.

1. How may cells make up the oligocytous intestine seen in Rhabitis *sp.?*______________

2. How may cells form the polycytous intestine in Mononchus *sp.?* ________________

3. What is a possible function of the sphaerocrystals of rhabditin seen in the intestinal cells of Rhabditis *sp.?* ____

____

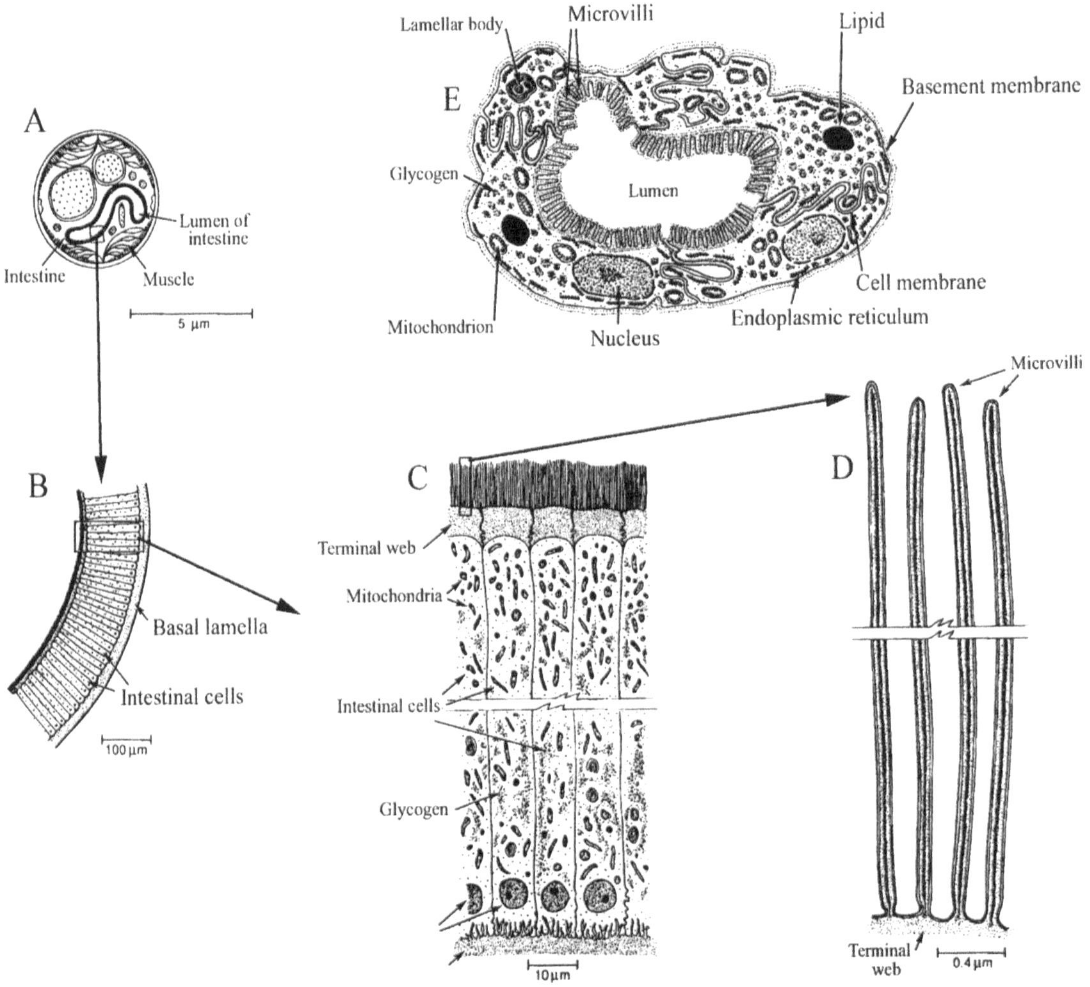

Fig. 15.1 Nematode intestine. A-D. Diagrams of the cross-section through *Ascaris suum* showing a progessinve increase in magnification (after Bird and Bird) E. Drawing of a cross-section through the four cells forming the circumference of *Trichinella spiralis* (after Bruce)

4. *What is the function of the intestinal mircrovilli?* ______________________

____

5. *What is the purpose of the anterior caecum in the male root-knot nematodes?* ______

____

____

Investigation 15

Posterior Gut

The posterior gut of female nematodes is a simple tube that connects the intestine to the anus. It is lined with cuticle and often has several rectal glands opening into it. The posterior intestinal cells form a simple intestino-rectal valve that is surrounded by a single sphincter muscle cell which keeps the valve closed. During defecation, this sphincter muscle relaxes and allows fecal matter to flow out of the intestine through the anal opening.

In the male, however, the intestine and reproductive system open into a common duct, the cloaca. The spicules and other accessory copulatory structures are contained within the cloaca.

Objectives of the Laboratory

Upon completion of this lab the student should be able to:

1. **Observe** the rectum and rectal glands of several female nematodes.
2. **Draw** and label the major parts of the posterior gut and accessory structures of various male nematodes.

Materials needed for this laboratory:

Nematode species:

Bunostomum sp.
Dorylaimus sp.
Meloidogyne sp. female
Rhabditis sp.
Diplogaster sp.
Hoplolaimus sp.
Pratylenchus sp.

For each student:

- Forceps
- Light microscope
- Nematode pick
- Slide storage box

For each pair of students:

- Acid fuchsin lactophenol
- Alcohol lamp
- Butyl acetate
- Cheesecloth
- 2% formalin
- Glass coverslips
- Glass microscope slides
- Immersion oil
- Matches
- Nail polish
- Slide labels
- Slide ringer
- Small paint brush
- Syracuse dishes
- Water

Work Sheet

I. Examine the rectum and the rectal glands of young females of *Rhabditis* sp.

II. Observe the 6 large rectal glands in young females of *Meloidogyne* sp. that have been stained with acid fuchsin lactophenol.

III. Observe the prerectum in females of *Dorylaimus* sp.

IV. Compare different types of male accessory structures.

A. *Diplogaster* sp., spicules free, one gubernaculum.

B. *Rhabditis* sp., spicules fused distally, one gubernaculum.

C. *Dorylaimus* sp., spicules free, 2 accessory pieces.

D. *Pratylenchus* sp., spicules free (tylenchoid type), one gubernaculum.

E. *Hoplolaimus* sp., spicules free (tylenchoid type), gubernaculum with titillae, capitulum on ventral cloacal wall.

F. *Bunostomum* sp., spicules very long, fused distally, gubernaculum lacking.

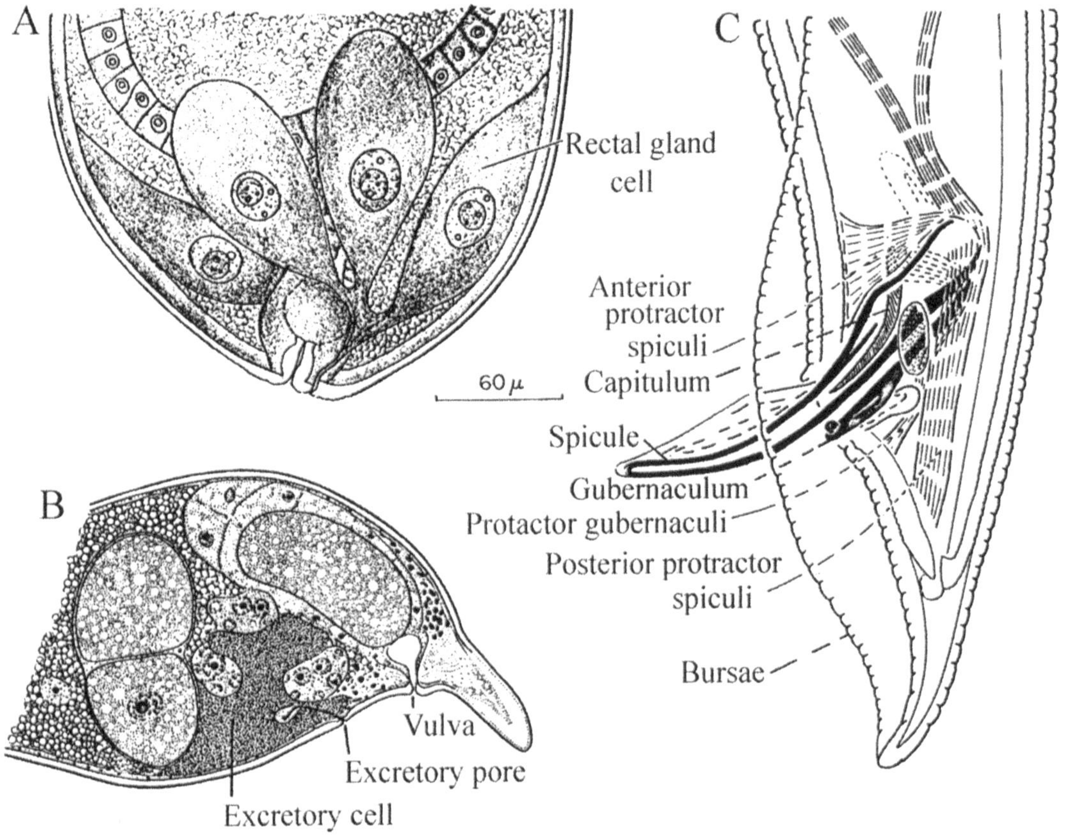

Fig. 15.1 A. Posterior region of a female root-knot nematode, *Meloidogyne incognita*, showing four of the six rectal gland cells B. Excretory system of *Tylenchulus semi-penetrans* (saggital section) C. Tail of a male lance nematode, *Hoplolaimus* sp.

1. *What is the purpose of the rectal glands in females of* Rhabditis *spp.?* ______________

2. *What is the purpose of the rectal glands in females of* Meloidogyne *spp.?* ______________

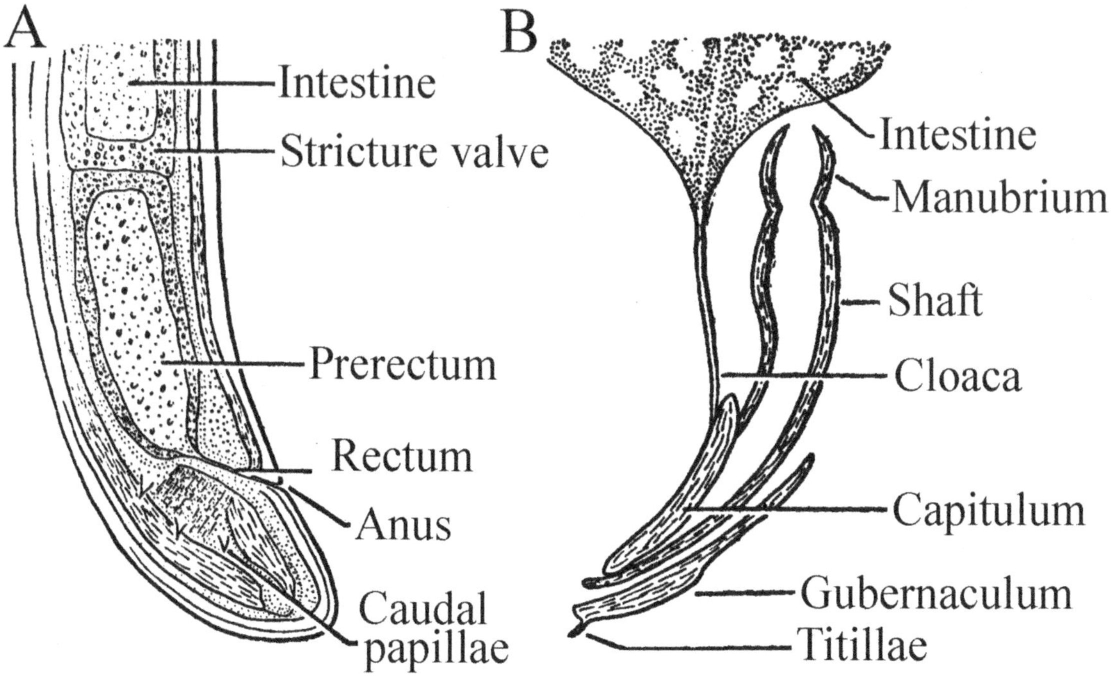

Fig. 15.2 Posterior gut. A. Female with anus and rectum B. Male with spicule and cloaca (after Caveness)

3. Discuss the importance of the six rectal glands in root-knot females. ______________

____

____

4. Where does the intestine open in the root-knot female? ______________________

____

5. Compare and contrast the various types of male accessory sructures. ____________

____

____

____

____

Investigation 16

Nervous System and Sensory Structures

Observation of the nervous system of nematodes is usually limited to the large nerve ring and associated ganglia that surround the esophagus in the anterior end of the body and various sensory structures.

Sensory structures are common to all nematodes. Some of them, such as the amphids, occur in all groups of nematodes, whereas the phasmids are present in members of the class Secernentea but absent in the class Adenophorea. Nematodes may possess eye spots, touch receptors, and chemoreceptors in the form of papillae, sensilla, setae, bristles, pores, warts, and other manifestations. The occurrence and form of many of these sensory structures are important taxonomic characters. Many, however, are difficult to observe. Patience and careful observations with constant focusing are often necessary in order to see many of these structures.

Objectives of the Laboratory

Upon completion of this lab the student should be able to:

1. **Observe** and **draw** the nerve ring and associated ganglia.
2. **Examine** labial papillae and cephalic and somatic setae.
3. **Describe** the morphology of several amphids.
4. **Examine** and **draw** deirids and phasmids.
5. **Draw** the supplementary organs of several males.

Materials needed for this laboratory:

Nematode species:

Anaplectus sp.
Anguina sp.
Ditylenchus sp.
Dorylaimus sp.
Draconema sp.
Helicotylenchus dihystera
Hoplolaimus galeatus
Monhystera sp.
Mononchus sp.
Rhabditis sp.
Scutellonema sp.
Tobrilus sp.
Trichodorus sp.
Tylenchorhynchus sp.

For each student:

Forceps
Nematode pick
Light microscope
Slide storage box

For each pair of students:

Alcohol lamp
Nail polish
Butyl acetate
Propionic orcein
Cheesecloth
Slide labels
2% formalin
Small paint brush
Glass coverslips
Syracuse dishes
Glass microscope slides
Immersion oil
Matches
Water

Work Sheet

I. Observe and draw the nerve ring and associated ganglia.

A. *Mononchus* sp. Also note the papillary nerves in the lip region.

B. *Helicotylenchus dihystera.* Stain specimens with propionic orcein. Nuclei of the ganglia around the nerve ring and in ventral chord stain red; however, the nerve ring should remain unstained. Also, observe the hemizonid anterior to the excretory pore.

II. Examine the labial papillae, cephalic setae, and the somatic setae.

A. *Tobrilus* sp. Observe the labial papillae and cephalic setae.

B. *Draconema cephalatum.* Examine the somatic setae and the loop-shaped amphids.

III. Describe the morphology of the amphids.

A. Compare the spiral type of the Chromadorida (marine nematode) with the circular type of the Monhsterida (*Monhystera* sp.) and the question mark type of Plectidae (*Plectus* sp.).

B. *Trichodorus* sp. Study the amphidial opening leading into the amphidial pouch filled with several sensilla.

IV. Examine the deirids.

A. *Ditylenchus* sp.

B. *Anguina* sp.

V. Study the phasmids.

A. *Tylenchorhynchus* sp. (small type).

B. *Scutellonema brachyurum* (large type scutellum on tail).

C. *Hoplolaimus galeatus* (one on each lateral side, 25-65% and 75-90% from anterior end).

VI. Draw the supplementary organs of the male.

A. *Dorylaimus* sp. has papilloid supplementary organs.

B. *Anaplectus* sp. has tuboid sclerotized preanal organs that terminate in glands.

1. Compare and contrast a phasmid with a scutellum. ____________________

____

____

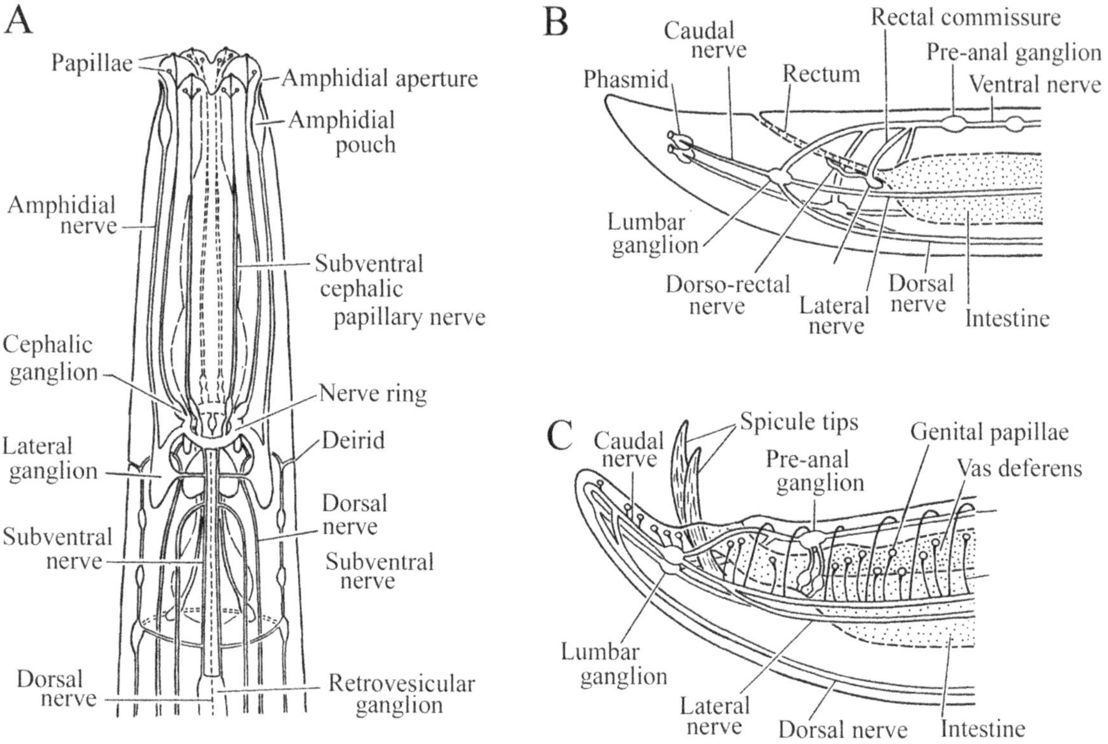

Fig. 16.1 A. Anterior part of *Rhaditis terricola* (after Hirschmann, modified after Chitwood and Chitwood) B. Posterior part of female nematode (after Hirschmann, modified after Crofton) C. Posterior part of male of *Ascaris* sp. (after Hirschmann, modified after Hyman)

2. *What is the probable function of labial papillae?* ______________________

____

3. *What is the theoretical function of the deirids?* ______________________

____

4. *What is the purpose of amphids?* ______________________

5. *Do all nematodes have amphids?* ______________________

6. *Why are the aphids paired organs?* ______________________

7. *What is the function of supplemental organs in the male?* ______________________

____

Investigation 17

Secretory/Excretory Systems

The secretory/excretory system is diverse in both form and function. In Adenophorean nematodes, the system is simple and glandular, usually consisting of a single glandular cell connected to a terminal duct. This normally elongated duct loops within the pseudocoelom and the glandular tissue. Usually, it is not lined with cuticle and, thus, difficult to see. In Secernentean systems, the morphology is more complex and diverse. The terminal duct is usually lined with cuticle and easier to discern. This system is composed of numerous cells and is characterized by long canals in the lateral chords. Variations in the secretory/excretory system occur from the loss of some of the component of the complete H-shaped system.

Objectives of the Laboratory

Upon completion of this lab the student should be able to:

1. **Identify** the secretory/excretory system of Adenophorean and Secernentean nematodes.
2. **Draw** and **label** the different types of nematode secretory/excretory systems.

Materials needed for this laboratory:

Nematode species:

Ascaris lumbricoides cross-sections
Meloidogne sp. females
Plectus sp.
Meloidogyne sp. males
Rhabditis sp.

For each student:

Forceps
Light microscope
Nematode pick
Slide storage box

For each pair of students:

Alcohol lamp
Butyl acetate
Cheesecloth
2% formalin
Glass coverslips
Glass microscope slides
Immersion oil
Matches
Nail polish
Slide labels
Slide ringer
Small paint brush
Syracuse dishes
Water

Work Sheet

I. Observe and make sketches:

A. *Meloidogyne* sp. males. Excretory pore located behind the nerve ring; terminal excretory duct long; lateral canal ending in sinus cell with large nucleus. Note the hemizonid anteriad of the excretory pore in the hypodermis.

B. *Meloidogyne* sp. females. The excretory pore is located far anteriad at the height of the stylet knobs. The terminal duct is long and strongly sclerotized.

C. *Plectus* sp. Cuticularly lined excretory duct makes 2 loops after entering single excretory gland cell.

D. *Ascaris* esophagus cross-section. One large excretory canal located in each of the lateral chords.

E. Live specimens of *Rhabditis* sp. in a temporary water mount. The excretory canals are located within the 2 lateral chords, and proceed anteriad and posteriad. Two ventral excretory gland cells are present, and a large sinus nucleus is in their vicinity. The terminal excretory duct is coiled and strongly sclerotized.

1. What is the function of the secretory/excretory system? ______________________

____

____

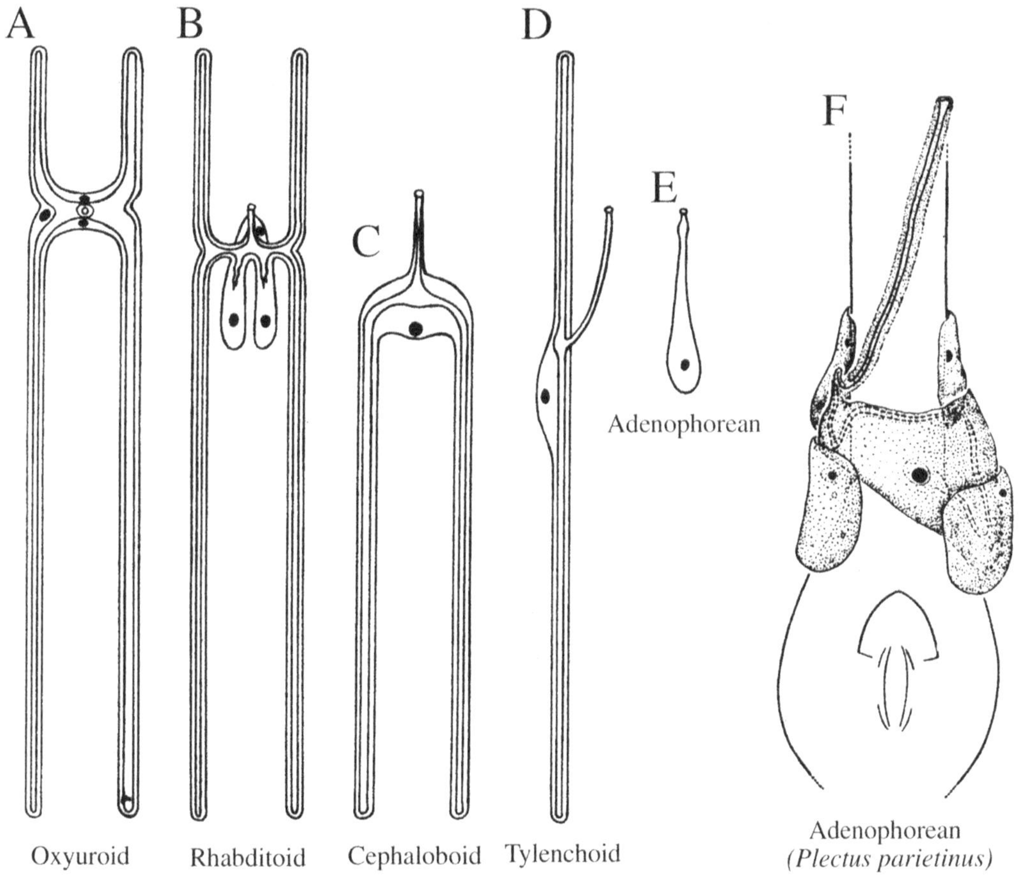

Fig. 17.1 Secretory/excretory systems of Secernentean and Adenophorean nematodes. A. Oxyuroid type B. Rhabditoid type C. Cephaloboid type D. Tylenchoid type E. Single ventral cell (Adenophorean type) F. Adenophorean type from *Plectus parietinus* (A-E after Hirschmann redrawn from Chitwood and Chitwood, F after Maggenti)

2. *Compare and contrast the Adenophorean secretory/excretory system with the Secernentean system.* ______________________________

____

____

____

____

Investigation 18

Reproductive Systems

The reproductive systems of nematodes vary in form and in detail. Basically, the gonads are tubular structures that are contained within the pseudocoelom. The female reproductive system can have one or two ovaries; Adenophorean nematodes ususally have one ovary, whereas Secernentean females usually have two ovaries. In males, there are usually two testes in Adenophorean species and one testis in Secernentea. Both sexes may have secondary sexual characters. Males may develop copulatory structures such as bursae, papillae, gubernaculi, and spicules.

Objectives of the Laboratory

Upon completion of this lab the student should be able to:

1. Describe a typical female reproductive system.
2. Draw and label a typical male reproductive system.
2. Outline variations in the morphology of female reproductive systems.
4. Draw and label several types of male reproductive systems.

Materials needed for this laboratory:

Nematode species:

Anaplectus sp.
Helicotylenchus sp.
Meloidogyne hapla
Pratylenchus sp.
Diplogaster sp.
Hoplolaimus sp.
Meloidogyne javanica
Rhabditis sp.

For each student:

Forceps
Light microscope
Nematode pick
Slide storage box

For each pair of students:

Alcohol lamp	Matches
Butyl acetate	Nail polish
Cheesecloth	Slide labels
2% formalin	Slide ringer
Glass coverslips	Small paint brush
Glass microscope slides	Syracuse dishes
Immersion oil	Water

Work Sheet

I. Study the female reproductive system of the following nematodes and make drawings:

A. *Diplogaster* sp. - Gonads didelphic (amphidelphic) and reflexed.

B. *Hoplolaimus* sp. - Gonads didelphic (amphidelphic); ovaries not reflexed. Observe the spermatotheca in line with ovary, and anterior uterus developed as tricolumella.

C. *Helicotylenchus* sp. - Gonads didelphic (amphidelphic); ovaries not reflexed. Observe dorsally-offset sperma totheca and tricolumella.

D. *Meloidogyne* sp. - Gonads didelphic (prodelphic); ovaries reflexed. Dissect young females in physiological salt solution and observe the gonads.

E. *Pratylenchus* sp. - Monodelphic gonad (prodelphic); ovary not reflexed. Observe the round spermatotheca and short postvulval uterine sac.

II. Study the male reproductive system of the following nematodes and make drawings:

A. *Anaplectus* sp. - Two testis (diorchic) extending in opposite directions.

B. *Rhabditis* sp. - Single (monorchic) reflexed testis; vas deferens with two ejaculatory glands.

C. *Hoplolaimus* sp. - Single (monorchic) outstretched testis.

D. *Meloidogyne hapla* - Two testes (diorchic) oriented parallel. In some specimens only one testis.

E. *Meloidogyne javanica* - Intersex with rudimentary vulva.

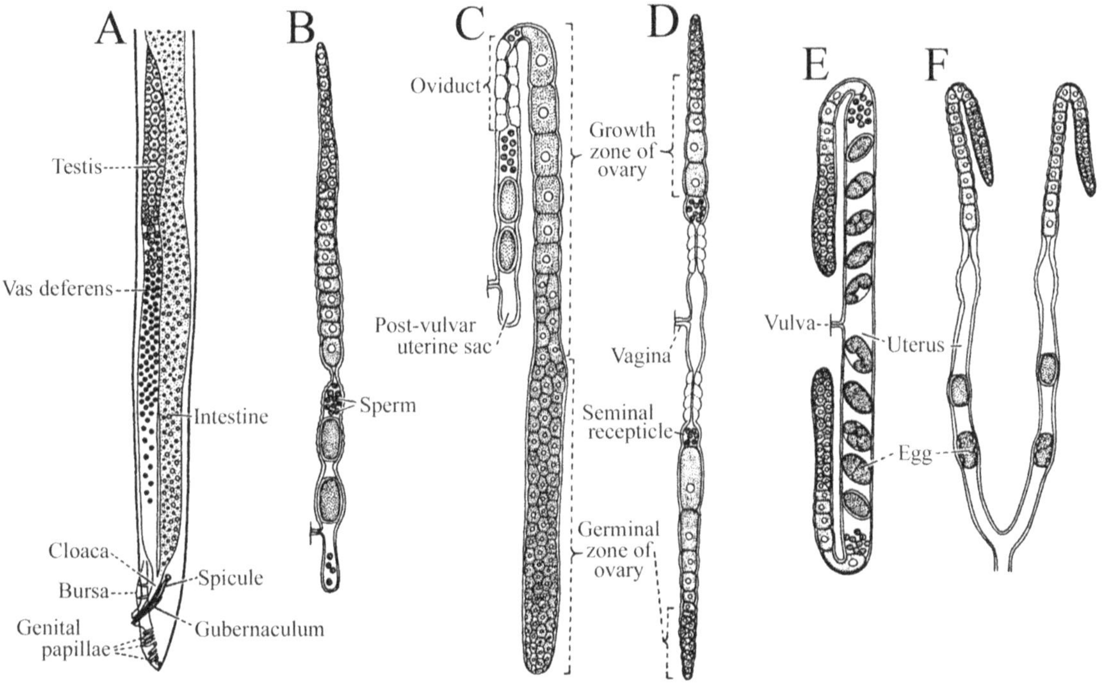

Fig. 18.1 Reproductive systems of nematodes. A. Monorchic male reproductive system of *Rhabditis* sp. B. Monodelphic, prodelphic outstretched ovary of *Ditylenchus* sp. C. Monodelphic, prodelphic reflexed ovary of *Panagrolaimus* sp. D. Didelphic, amphidelphic branches opposed ovaries of *Tylenchorhynchus* sp. E. Didelphic, amphidelphic branches opposed reflexed ovaries of *Rhabditis* sp. F. Didelphic, amphidelphic, prodelphic reflexed ovaries of *Meloidogyne* sp. (after Hirschmann)

1. Discuss a nematode's reproductive potential in relation to the number of eggs it can lay and the rapidity of its life cycle. ______________________________

____

____

____

____

____

____

Investigation 19

Female Reproductive System

The reproductive systems of nematodes vary in form and in detail. Basically, the gonad is a tubular structure that is contained within the pseudocoelom where it may be outstretched or folded one or more times. It is composed of two basic parts: the ovary and the gonoduct. The ovary proper is usually divided into three zones: the germinative, growth, and ripening zones. The gonoduct may have specialized regions as well, including the spermatheca, oviduct, and uterus. The female reproductive system can have one or two ovaries; Adenophorean nematodes ususally have one ovary, whereas females of Secernentea usually have two ovaries.

Objectives of the Laboratory

Upon completion of this lab the student should be able to:

1. **Describe** a typical female reproductive system.
2. **Outline** variations in the morphology of female reproductive systems.

Materials needed for this laboratory:

Nematode species:

Diplogaster sp.
Helicotylenchus sp.
Meloidogyne sp.
Pratylenchus sp.
Dorylaimus sp.
Hoplolaimus sp.
Panagrellus sp.

For each student:

Forceps
Light microscope
Nematode pick
Slide storage box

For each pair of students:

Alcohol lamp	Matches
Butyl acetate	Nail polish
Cheesecloth	Slide labels
2% formalin	Slide ringer
Glass coverslips	Small paint brush
Glass microscope slides	Syracuse dishes
Immersion oil	Water

Work Sheet

I. Study the female reproductive system of the following nematodes and make drawings.

A. *Diplogaster* sp. - Gonads didelphic (amphidelphic); ovaries reflexed with extended germinal and growth zones.

B. *Dorylaimus* sp. - Gonads didelphic (amphidelphic); ovaries reflexed with short germinal and growth zones.

C. *Hoplolaimus* sp. - Gonads didelphic (amphidelphic); ovaries not reflexed. Observe the spermatotheca in line with ovary, and anterior uterus developed as tricolumella.

D. *Helicotylenchus* sp. - Gonads didelphic (amphidelphic); ovaries not reflexed. Observe dorsally-offset spermatotheca and tricolumella.

E. *Meloidogyne* sp. - Gonads didelphic (prodelphic); ovaries reflexed. Dissect young females in physiological salt solution and observe the gonads.

F. *Panagrellus* sp. - Monodelphic gonad (prodelphic); ovary reflexed. Observe the postvulval uterine sac.

G. *Pratylenchus* sp. - Monodelphic gonad (prodelphic); ovary not reflexed. Observe the round spermatotheca and short postvulval uterine sac.

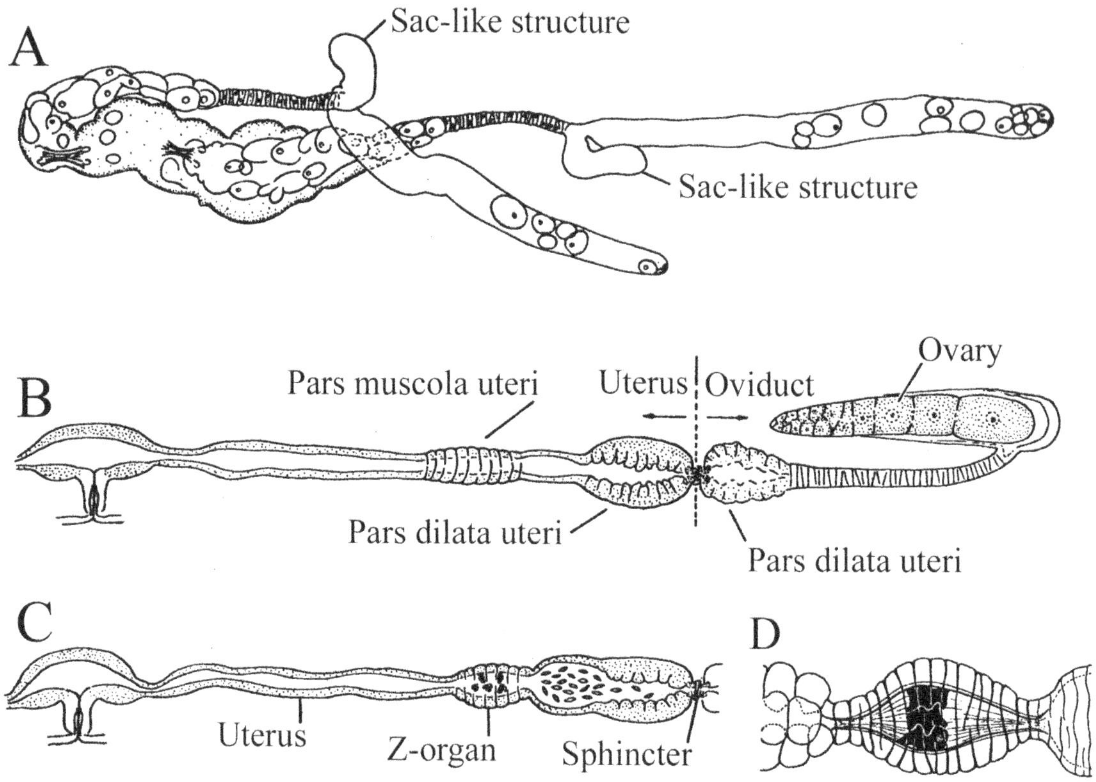

Fig. 19.1 Female reproductive systems of Adenophorean nematodes. A. Reproductive system of young female *Longidorus elongatus* (after Williams) B. Anterior gonad of a Dorylaimid nematode (after Coomans) C. Uterus of a Dorylaimid nematode (after Coomans) D. Z-organ of *Xiphinema ifacolum* (after Luc)

1. *Compare and contrast the ovaries of* Diplogaster *sp. and* Meloidogyne *sp.* ________

____

____

____

2. *Why are the ovaries of some females reflexed?* ________

____

____

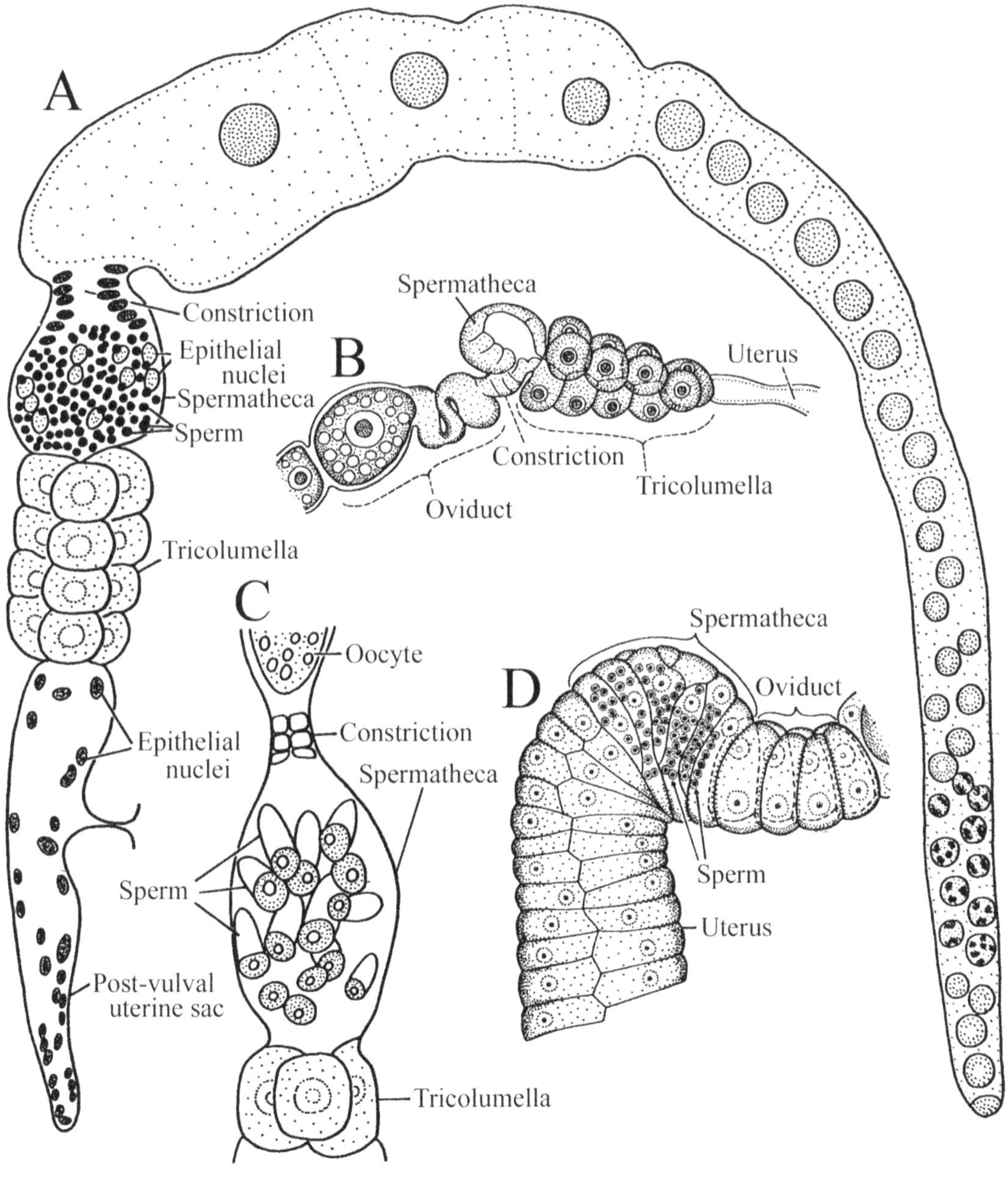

Fig. 19.2 Female reproductive systems of Tylenchida. A. Female gonad of *Pratylenchus coffeae* (after Roman and Hirschmann) B. Portion of the anterior gonad of *Helicotylenchus dihystera* (after Hirschmann and Triantaphyllou) C. Spermatheca of *Pratylenchus vulnus* (after Roman and Hirschmann) D. Oviduct, spermatheca and part of uterus of *Heterodera glycines* (after Triantaphyllou and Hirschmann)

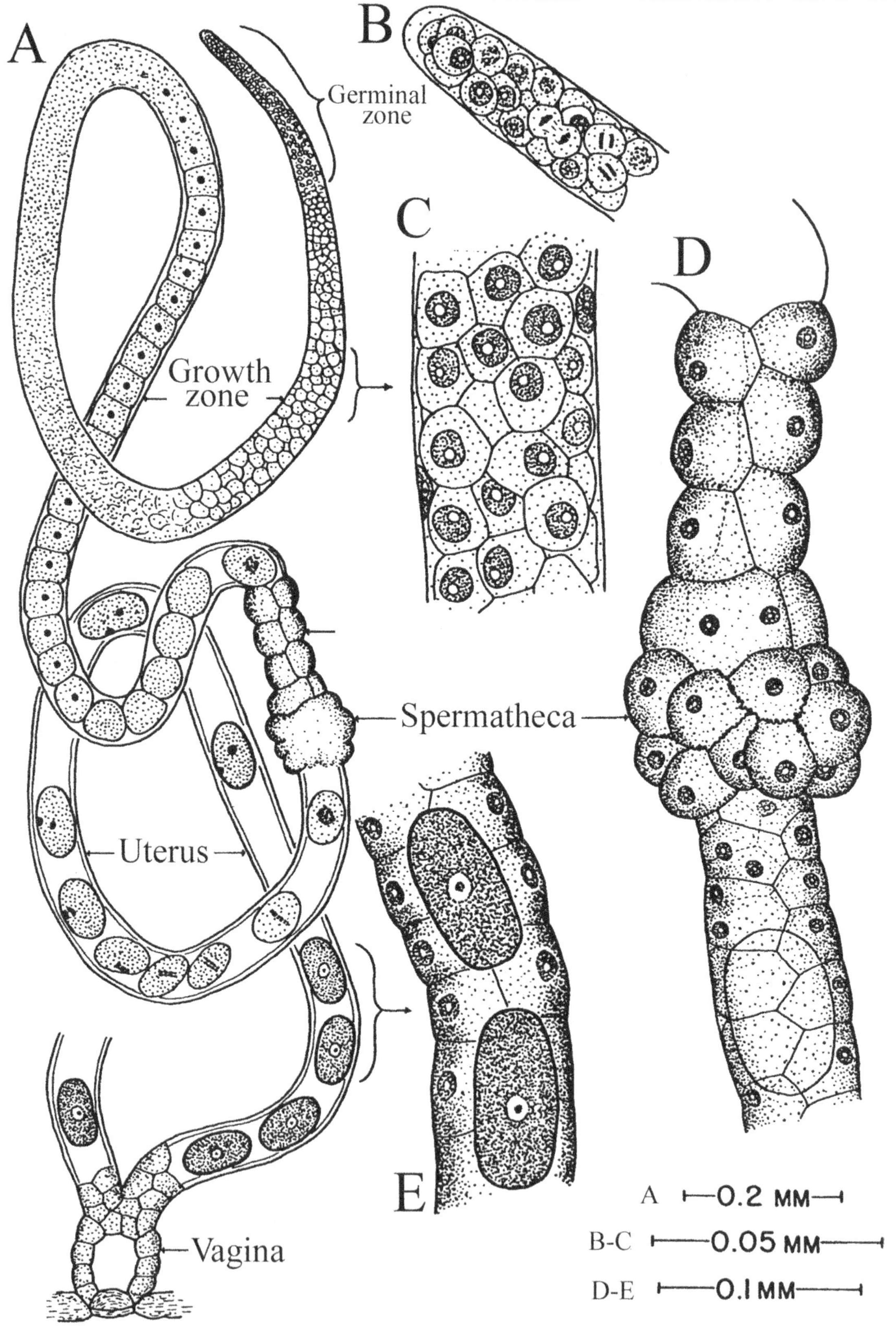

Fig. 19.2 Female reproductive system of *Meloidogyne javanica*. A. One of two gonads showing the various regions B. Distal end of the germinal zone of the ovary C. Growth zone of the ovary D. Oviduct and spermatheca E. Lower region of the uterus with an egg (after Triantaphyllou)

Investigation 20

Male Reproductive System

The reproductive systems of nematodes vary in form and in detail. Basically, the gonad is a tubular structure that is contained within the pseudocoelom where it may be outstretched or folded back. There are usually two testes in Adenophorean males and one testis in Secernentean males. The testis is composed of two basic parts: the testis and the gonoduct. Usually the testis proper can be divided into two regions: the germinal and growth zones. The gonoduct usually has a spermatheca (seminal vesicle) and vas defferens, but also ejaculatory and cloacal glands can also be present. Male copulatory structures may include bursae, spicules, gubernaculi, papillae and other characters.

Objectives of the Laboratory

Upon completion of this lab the student should be able to:

1. **Describe** a typical male reproductive system.
2. **Outline** variations in the morphology of male reproductive systems.

Materials needed for this laboratory:

Nematode species:

Dorylaimus sp.
Meloidogyne sp.
Pratylenchus sp.
Hoplolaimus sp.
Parascaris equorim

For each student:

Forceps
Light microscope
Nematode pick
Slide storage box

For each pair of students:

- Alcohol lamp
- Butyl acetate
- Cheesecloth
- 2% formalin
- Glass coverslips
- Glass microscope slides
- Immersion oil
- Matches
- Nail polish
- Slide labels
- Slide ringer
- Small paint brush
- Syracuse dishes
- Water

Work Sheet

I. Study the male reproductive system of the following nematodes and make drawings:

A. *Rhabditis* sp. Single (monorchic) reflexed testis. Vas deferens with two ejaculatory glands, testis with short germinal and growth zones.

B. *Hoplolaimus* sp. Single (monorchic) outstretched testis.

C. *Meloidogyne javanica* Two testes (diorchic) oriented parallel; in some specimens only one testis.

D. *Dorylaimus* sp. Two testes (diorchic) oriented opposite.

E. Observe the sperm of *Rhabditis* sp. and *Hoplolaimus* sp. and compare them with those of *Parascaris equorum*.

F. Study the development of the reproductive system during post-embryogenesis in juvenile stages of *Meloidogyne* sp.

1. Compare and contrast the testis of Rhabditis *sp. with that of* Hoplolaimus *sp.* _______

____

____

____

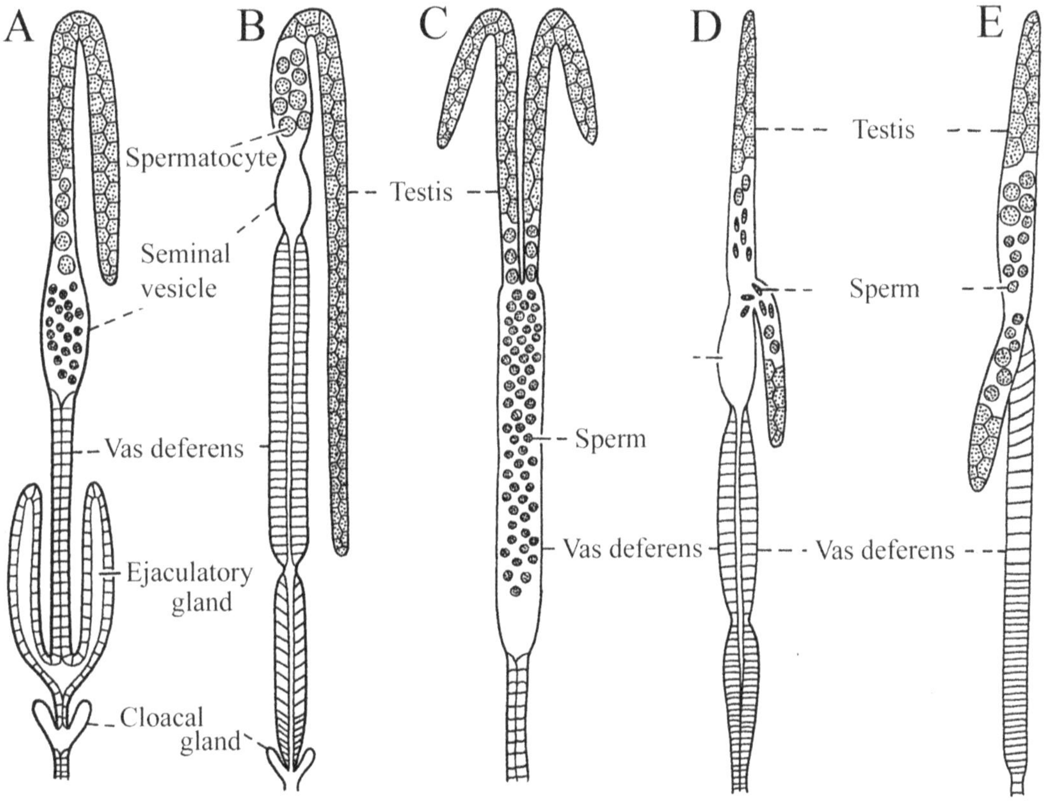

Fig. 20.1 Reproductive systems of male nematodes. A. Monochic, reflexed testis of *Rhabditis strongyloides* B. Monorchic, reflexed testis of *Spironura affine* C. Diorhic reflexed testes of *Meloidogyne hapla*, directed anteriad D. Diorchic, outstretched, opposed testes of *Enoplus communis* E. Diorchic outstretched, opposed testes of *Anaplectus granulosus* (after Hirschmann)

2. *Why do some males of* Meloidogyne javanica *have two testes, whereas others have just one?* ____________________

____

3. *Compare and contrast the sperm of* Rhabditis *sp. and* Parascaris equorum. ____________________

____

____

Investigation 21

Eggs

Nematode eggs are usually rounded, oval, or elliptical in shape, and they are surrounded by three layers. The innermost layer is the lipid layer; the true shell is a chitinous layer which is secreted by the egg itself and often covered with a vitelline membrane. The outermost layer is a proteinacious uterine layer which is secreted by the uterine wall. This outer layer is absent in some nematode eggs. In general, egg morphology is related to the ecology of the nematode. Some aquatic forms have small spines or hooks, and eggs of some animal parasites may have specialized structures such as branched polar cords, polar or equatorial filaments, mammilations, or opercula.

Objectives of the Laboratory

Upon completion of this lab the student should be able to:

1. **Describe** a typical plant-parastic nematode egg.
2. **Observe** the mammilations on the surface of a typical anima-parasitic nematode.

Materials needed for this laboratory:

Nematode species:

Belonolaimus sp.
Meloidogyne sp.
Pratylenchus sp.
Hoplolaimus sp.
Parascaris equorim

For each student:

Forceps
Light microscope
Nematode pick
Slide storage box

For each pair of students:

- Alcohol lamp
- Butyl acetate
- Cheesecloth
- 2% formalin
- Glass coverslips
- Glass microscope slides
- Immersion oil
- Matches
- Nail polish
- Slide labels
- Slide ringer
- Small paint brush
- Syracuse dishes
- Water

Work Sheet

I. Observe and draw the morphology of the eggs of several plant-parasitic nematodes:

A. *Belonolaimus* sp

B. *Hoplolaimus* sp.

C. *Meloidogyne* sp.

D. *Pratylenchus* sp.

II. Observe and draw the mammilations on the surface of the eggs of *Parascaris equorim.*

1. Explain the relationship between the egg morphology and nematode ecology. ______

____

____

____

____

2. Why are most nematode eggs elliptical in shape? ________________

____

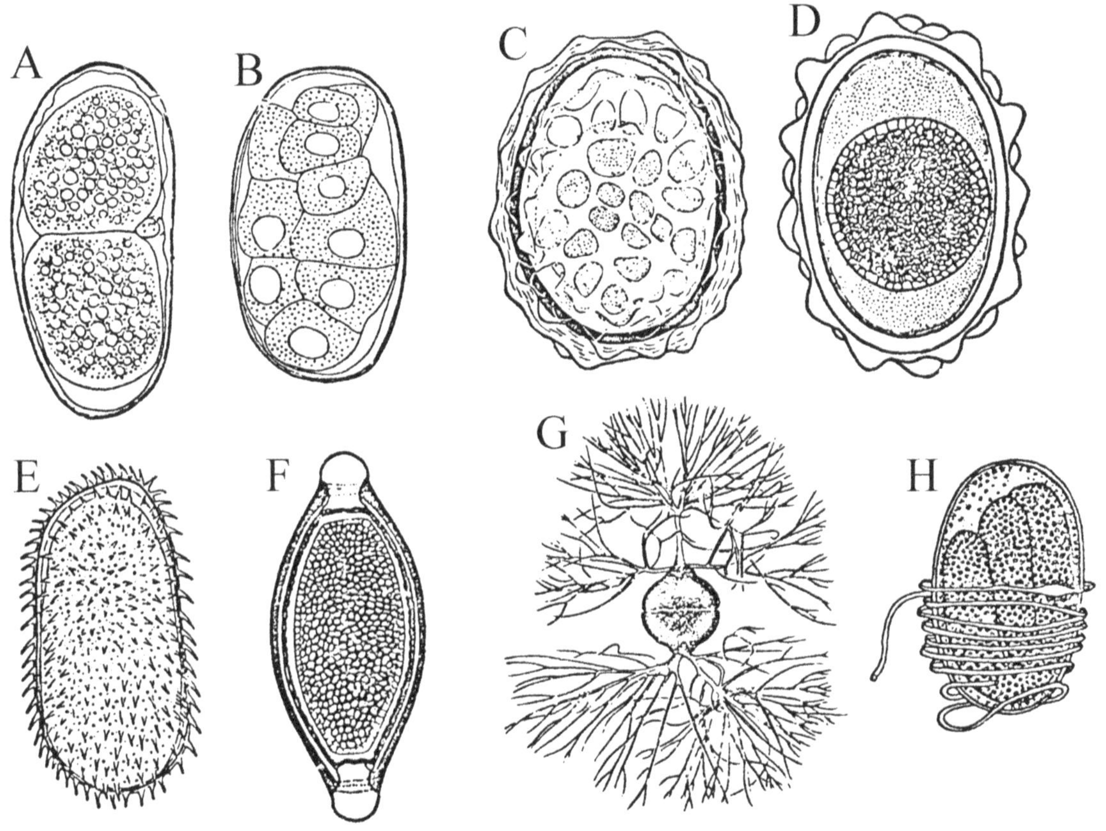

Fig. 21.1 Eggs of various nematodes. A. *Meloidogyne hapla* B. *Rhabditis strongyloides* C-D. *Ascaris lumbricodies* with mammilation E. *Anaplectus granulosus* with spination F. *Trichuris leporis* with distinct opercula G. *Mermis subnigrescens* with byssi H. *Pseudonymus* sp. with equatorial filaments

3. *Why do many aquatic nematode eggs have small spines and hooks?* ______________

____

____

____

4. *What is the purpose of the byssi on the eggs of* Mermis subnigrescens*?* ____________

____

____

Investigation 22

Preparation of Permanent Mounts

Long-term collection of nematodes for voucher and type specimens requires special preparation and mounting techniques. Permanent mounts of nematodes that are well prepared and mounted in glycerine may be preserved almost indefinitely. The addition of a general stain such as picric acid or cotton blue can enhance the definition of certain structures. For plant-parasitic nematodes, the most common fault of permanent mounting techniques is the dissolution of the stylet shaft and knobs. Likewise, saccate females are more difficult to preserve and mount because of their awkward shape.

Most permanent mounting techniques are either fast or slow methods. Fast methods usually require more handling of the specimens but can be as fast as one hour. Slow methods generally require less handling, but they may take several days or weeks to perform.

Objectives of the Laboratory

Upon completion of this lab the student should be able to:

1. **Make** a permanent mount of a nematode.
2. **Outline** the procedures for making permanent mounts of nematodes.

Materials needed for this laboratory:

Nematode species:

Various nematode species of the instructor's or student's choice

For each student:

Forceps
Light microscope
Nematode pick
Slide storage box

For each pair of students:

Adhesive tape	Glass microscope slides
Alcohol lamp	Glass fibers
Baermann funnels	Glycerin
BEEM capsules	Hole puncher
BPI dishes	Immersion oil
Butyl acetate	Matches
Cheesecloth	Nail polish
Cobb slide mounts	Petri dishes
Dessicator	Pipettes
FAA	Slide labels
FP 4:1	Slide ringer
2% formalin	Small paint brush
Glass coverslips	Water

Work Sheet

I. Practice making permanent mounts according to Seinhorst's Rapid Method starting with nematodes that have already been processed into glycerin.

A. Use the provided fixed nematodes and start with step D of the method. Complete the remaining steps before the next laboratory.

B. Use the provided living nematodes and kill them with hot FP 4:1. Transfer the nematodes to 2% formalin and fix for 24-48 h; proceed with steps D through G.

II. Prepare permanent mounts according to a method of your choice from those outlined on the following pages.

III. Continue making semi-permanent mounts in 2% formalin or glutaraldehyde of five different genera of your choice including either *Meloidogyne* sp., *Rotylenchulus* sp., *Globodera* sp., or *Heterodera* sp. Try at least one rapid method and one slow method.

PREPARATION OF PERMANENT MOUNTS OF SPECIMENS PROVIDED BY YOUR INSTRUCTOR (Use Cobb slides)

A. Wash 25 mm square coverslips (#1 thickness) with alcohol.

B. Attach coverslip to glass slide with small pieces of scotch tape.

C. Add small drop of desiccated glycerin to center of square coverslip.

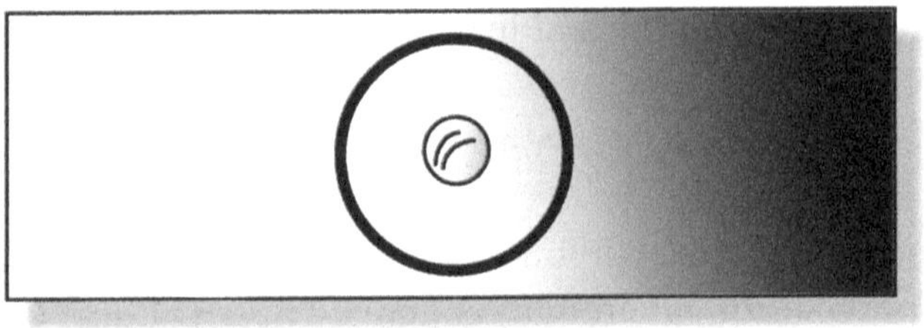

D. Place 5-10 nematodes at bottom of glycerin drop. Select only specimens of the same genus.

E. Place 3 short glass fibers (angel hair) in glycerin drop as coverslip supports. These fibers should be slightly smaller in diameter than the nematodes.

F. Wash 18 mm round coverslips (#1 thickness) with alcohol.

G. Heat round coverslip gently over alcohol flame and place over glycerin drop with curved forceps. It is important to center the coverslip properly!

H. Tack round coverslip at 2-3 points with nail polish.

I. After nail polish has dried (15-20 min), remove excess glycerin with small filter paper strips and ring coverslip completely with a thick seal of nail polish.

J. Examine mounts with the oil immersion objective and select good nematode preparations.

K. Remove good coverslip mounts from glass slides and place into Cobb slides.

L. Label Cobb slides with the following information:

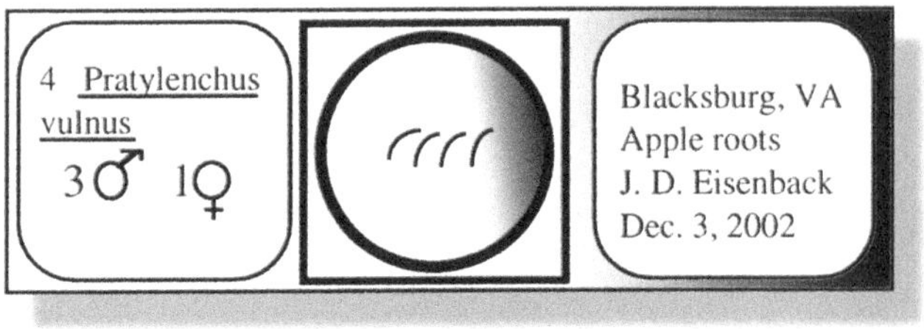

Instructions for Making Permanent Nematode Mounts in Glycerin

Thorne's slow method

1. Relax specimens by gentle heat in a drop of water on a slide.

2. Transfer nematodes to a vial with FAA or 2% formalin and fix for 24-48 h.

3. Transfer specimens to a solution of 1.25% glycerin in 20% alcohol (add trace of copper sulfate to prevent fungal growth) contained in a BPI dish or round-bottom glass dish.

4. Place container with nematodes in partially closed petri dish and transfer the latter to a small airtight desiccator containing calcium carbonate as desiccating agent. Leave for 4-6 weeks at 30-35°C.

5. After complete evaporation of the alcohol, place container with nematodes in calcium chloride desiccator for 2-3 days.

6. Mount nematodes in anhydrous glycerin on Cobb aluminum slides.

Slow Method Using Cold Glutaraldehyde as Fixative

1. Pick 20-30 live nematodes into 15 drops of tap water in a BPI dish.

2. Place BPI dish in petri dish lines with a moist piece of filter paper and chill nematodes in the refrigerator at 8°C for at least 30 min.

3. Gradually fix nematodes by adding 3 drops of cold 4% glutaraldehyde every 10-15 min until 20 drops of fixative have been added (final fixative concentration: 2% glutaraldehyde).

4. Keep the specimens in the fixative for an additional 48-72 h in the cold.

5. Wash nematodes several times in cold sodium cacodylate buffer, then pick them into buffer contained in a round-bottom dish, and keep for 24 h in the cold.

6. Bring buffer to room temperature. Place nematodes in a solution of 1.5% glycerin in water of room temperature (add trace of copper sulfate to prevent fungal growth) contained in a round-bottom glass dish covered with perforated parafilm.

7. Infiltrate the nematodes slowly at room temperature or in an oven at 34-38°C regulating speed of evaporation by size of parafilm holes.

8. After complete evaporation of alcohol, place round-bottom dish with nematodes in calcium chloride desiccator for 2-3 days.

9. Mount specimens in desiccated glycerin on Cobb slides.

Caution: Glutaraldehyde and sodium cacodylate buffer are very toxic substances and should be handled with extreme caution only under a fume hood.

Modified Baker's rapid method for pyriform females of *Meloidogyne* and *Heterodera*

1. Fix female nematodes in FA 4:10 (formalin 10 ml, glacial acetic acid 10 ml, distilled water 80 ml) or 2% formalin in closed vials for 12-24 h.

2. Transfer nematodes to a BEEM capsule chamber* or to a specially constructed glass chamber (3 nested glass rings with cheese cloth covers) and clear in hot (60°C, near boiling) lactophenol contained in a 10 ml glass weighing jar for 3-5 min on a hot plate.
3. Process nematodes in a BEEM Capsule chamber through a series of glycerin-lactophenol solutions contained in 10 ml covered glass weighing jars. Leave 10 min at each step in an oven at 60°C.

Solution	1	2	3	4	5
Glycerin	50	70	80	90	100
Lactophenol	50	30	20	10	0

4. Transfer nematodes from solution 5 (100% glycerin) to a calcium chloride desiccator and leave for 2-3 days.

5. Mount specimens in anhydrous glycerin on Cobb aluminum slides.

* *BEEM® capsule chamber: Sever cap of capsule. With a hole-puncher 4 mm in diameter, make 2 holes in the cap which is used for the bottom of the chamber. Cut off tapered end of capsule and discard. Save hollow cylinder. Cut a small, 20-mm square or circular piece of nylon screen, place it over one end of cylinder and snap bottom cap in place. Place this chamber in Stendor dish or similar glass container with close-fitting lid.*

Seinhorst's Rapid Method

1. Transfer 20-30 live nematodes to a small drop of water in a Syracuse dish.

2. Pipette about 5 ml of hot FP 4:1 fixative quickly over the drop in the dish. Preparation of hot fixative: Fill a test tube half with FP 4:1 and place in a beaker filled with water. Heat the water to boiling on a hot plate.

3. Transfer specimens to a small vial filled with cold 2% formalin and fix for 1-2 days.

4. Pick specimens to a solution of 1% glycerin in 20% alcohol contained in a small round-bottom glass dish.

5. Place dish in a closed desiccator containing about 1/10 of its volume of 96% alcohol, and leave dish in this saturated atmosphere for at least 12 h in an oven at 35-40°C.

6. Decrease volume if necessary, then top up dish with a solution of 5% glycerin in 96% alcohol and place dish in a glass container which should be partially closed to allow slow evaporation. Maintain for at least 8 h at 40°C, or preferably longer, until all alcohol has evaporated and the nematodes are in pure glycerin.

7. Transfer nematodes to calcium chloride desiccator and leave for 2-3 days.

8. Mount nematodes in anhydrous glycerin on Cobb aluminum slides.

Three Slow Methods Recommended for Making Permanent Mounts of Plant-Parasitic Nematodes in Glycerin

Method A

1. Transfer 20-50 live nematodes to a small drop of water in a Syracuse dish.

2. Pour about 5 ml of hot FP 4:1 fixative quickly over the drop in the dish.
   Preparation of hot fixative: Fill a test tube half with FP 4:1 and place in 250 ml beaker filled one half of its volume with water. Heat the water to boiling on a hot plate.

3. Transfer specimens to small vial filled with cold FAA and fix for 1-3 days.

4. Pick specimens to a solution of 1.5% glycerin in 20% alcohol con in a small round-bottom glass dish. Fill dish almost to rim with alcohol-glycerin solution and cover tightly with parafilm. Puncture a few small holes in parafilm and transfer dish to oven at 35-40°C. Regulate slow evaporation of alcohol by size of parafilm holes. This process should take at least 2 weeks.

5. After complete evaporation of alcohol, place container with nematodes in calcium chloride desiccator for 2-3 days.

6. Mount nematodes in anhydrous glycerin on Cobb aluminum slides.

Method B

1. Transfer 20-50 live nematodes to centrifuge tube filled with 5 ml water at room temperature. After nematodes have settled, add 5 ml boiling 10% formalin (prepare hot 10% formalin as in A.2), and immediately shake tube to mix contents properly.

2. After nematodes have settled again, remove some of the supernatant fixative and transfer rest of fixative (now 5% formalin) with nematodes to small capped vial. Fix for up to 1 week in 5% formalin.

3. Pick specimens to a solution of 1.5% glycerin in water (add trace of copper sulfate to prevent fungal growth) contained in a small round-bottom glass dish. Fill dish almost to rim with water-glycerin solution and cover tightly with parafilm. Puncture small holes in parafilm to facilitate slow evaporation of water at room temperature. This process should take from 2-4 weeks.

4. After complete evaporation of alcohol, place container with nematodes in calcium chloride desiccator for 2-3 days.

5. Mount nematodes in anhydrous glycerin on Cobb aluminum slides.

Method C

1. Pick 20-50 live nematodes and wash them 3 times in tap water. Pick specimens into 15 drops of tap water in a BPI dish.

2. Place BPI dish in closed petri dish lines with a moist piece of filter paper, and chill nematodes in the refrigerator at 8° C for at least 30 min (nematodes should be completely straight).

3. Gradually fix nematodes by adding 3 drops of cold (8° C) 4% glutaraldehyde (buffered with sodium-cacodylate) every 10-15 min until 15 drops of fixative have been added (final fixative concentration: 2% glutaraldehyde).

4. Keep the specimens in the fixative for an additional 2-3 days in the cold.

5. Transfer the nematodes to a special processing chamber made from a BEEM capsule.* Wash the specimens in cold sodium-cacodylate buffer 5 times at 15 min intervals (can keep in buffer overnight in the cold).

6. Remove specimens from refrigerator and allow to attain room temperature. Then dehydrate with an alcohol series of 10, 20, 35, 50, 65, 80, 95 and 100%, 15 min at each step at room temperature (repeat 3 times in 100% alcohol).

7. Transfer nematodes to alcohol glycerin (90 ml 100% alcohol + 10 ml glycerin) in a round bottom glass dish. Seal dish with parafilm and punch a few small holes in film.

8. Infiltrate nematodes slowly in an oven at 34-38°C, regulating speed of evaporation by size of parafilm holes. This process should take at least 2 weeks.

9. After complete evaporation of alcohol, place round-bottom dish with nematodes in calcium chloride desiccator for 2-3 days.

10. Mount nematodes in anhydrous glycerin on Cobb aluminum slides.

* *BEEM® capsule chamber: Sever cap of capsule. With a hole-puncher 4 mm in diameter, make 2 holes in the cap which is used for the bottom of the chamber. Cut off tapered end of capsule and discard. Save hollow cylinder. Cut a small, 20-mm square or circular piece of nylon screen, place it over one end of cylinder and snap bottom cap in place. Place this chamber in Stendor dish or similar glass container with close-fitting lid.*

Investigation 23

Use of Nematological Keys

Identification of nematodes to genus and below can be difficult and time consuming. (Dicotomous and tabular keys are extremely valuable for the beginning student.) In order to utilize a key effectively, the user must have a working knowlege of nematode morphology and have a good understanding of the various character states that are used in making keys.

However, no matter how good the key, identification of an unknown nematode eventually requires a detailed study of original species descriptions, and perhaps even type or voucher specimens. In order for a correct identification to be made, the specimens have to be plentiful, in good physiological condition, well preserved, properly mounted, and adequate keys be available.

Objectives of the Laboratory

Upon completion of this lab the student should be able to:

1. **Use** a nematological key to identify a plant-parasitic nematode to genus.
2. **Key** a plant-parasitic nematode to species using a published key. (See appendix 6.)

Materials needed for this laboratory:

Nematode species:

Various plant-parasitic nematodes.

For each student:

Forceps
Light microscope
Nematode pick
Slide storage box

For each pair of students:

Alcohol lamp	Matches
Butyl acetate	Nail polish
Cheesecloth	Slide labels
2% formalin	Slide ringer
Glass coverslips	Small paint brush
Glass microscope slides	Syracuse dishes
Immersion oil	Water

Work Sheet

I. Prepare semi-permanent formalin mounts (2% formalin) of the nematodes provided.

A. Use nail polish rings as supports, square or round coverslips for mounting and seal the mounts with nail polish.

B. Select adult nematodes only.

II. Identify the nematodes to genera and, with the help of published keys, identify one of the nematodes to species.

Key to the most common genera of plant-parasitic nematodes.

1. Stylet absent...nonparasitic nematodes
1'. Stylet present: Mainly plant-parasitic nematodes...2

2. Two-part esophagus, median bulb absent, odontostyle;.......................Dorylaimida......3
2'. Three-part esophagus, distinct metacorpus, stomatostyle;.................Tylenchida..........5

3. Stylet long, straight; body long, slender; very large nematodes; ....Longidoridae......4
3'. Stylet short, curved, weak; body short, thick;.............Trichodoridae..........*Trichodorus*
Paratrichodorus

4. Stylet with flanges, guiding ring near base...*Xiphinema*
4'. Stylet without flanges, guiding ring near apex..*Longidorus*

5. Metacorpus more than 3/4 body width; DGO in metacorpus;........... Aphelenchina......6
5'. Metacorpus less than 3/4 body width; DGO in procorpus; ....Tylenchina......7

6. Female tail blunt, male without bursa; Aphelenchidae...............................*Aphelenchus*
6'. Female tail conoid with mucro, male without bursa; ....
....Aphelenchoidae.........*Aphelenchoides*

7. Procorpus separated from metacorpus...8
7'. Procorpus fused with metacorpus; Criconematidae..21

8. Female vermiform; Tylenchoidea...9
8'. Female swollen; Heteroderoidea..19

9. Stylet very long; head region distinctly rounded; ....Belonolaimidae.......10
9'. Stylet moderately long to short..11

10. Tail rounded..*Belonolaimus*
10'. Tail pointed...*Dolichodorus*

11. Stylet weak; ....Tylenchidae.........12
11'. Stylet strong;...13

12. Tail subacute, pointed; vulva in posterior 1/3...*Ditylenchus*
12'. Tail filiform, more than 6x body width at anus...*Tylenchus*

13. Vulva in middle of body; ....Tylenchorhynchidae...*Tylenchorhynchus*
13'. Vulva in posterior of body...14

14. Stylet strong; 2 ovaries; female tail less than BWA;......................Hoploplaiminae.....15
14’. Stylet relatively strong; head flattened;...Pratylenchinae....18

15. Scutellae...16
15’. Phasmids...17

16. Anterior and posterior to vulva...*Hoplolaimus*
16’. Both in anal region..*Scutellonema*

17. Ventral overlap; DGO > 1/4 stylet length...*Helicotylenchus*
17’. Dorsal overlap; DGO < 1/4 stylet length...*Rotylenchus*

18. One ovary; vulva posterior...*Pratylenchus*
18’. Two ovaries; vulva medial..*Radopholus*

19. Body white, no cyst, eggs deposited in a matrix;........Meloidogynidae....*Meloidogyne*
19’. Body brown hard cyst, most eggs retained;....................Heteroderidae....................20

20. Round cyst...*Globodera*
20’. Lemon shaped cyst...*Heterodera*

21. With a sheath; Hemicycliophoridae..*Hemicycliophora*
21’. Without a sheath...22

22. Female posterior swollen, ex. pore posterior; on roots;....Tylenchulidae...*Tylenchulus*
22’. Female posterior not swollen...23

23. Body short, thick; apparent segmentation; ....Criconematidae.......24
23’. Female small, slender, vulva posterior, no stylet in other stages............*Paratylenchus*

24. Cuticular annulations rounded..*Mesocriconema*
24’. Cuticular annulations serrated..*Criconema*

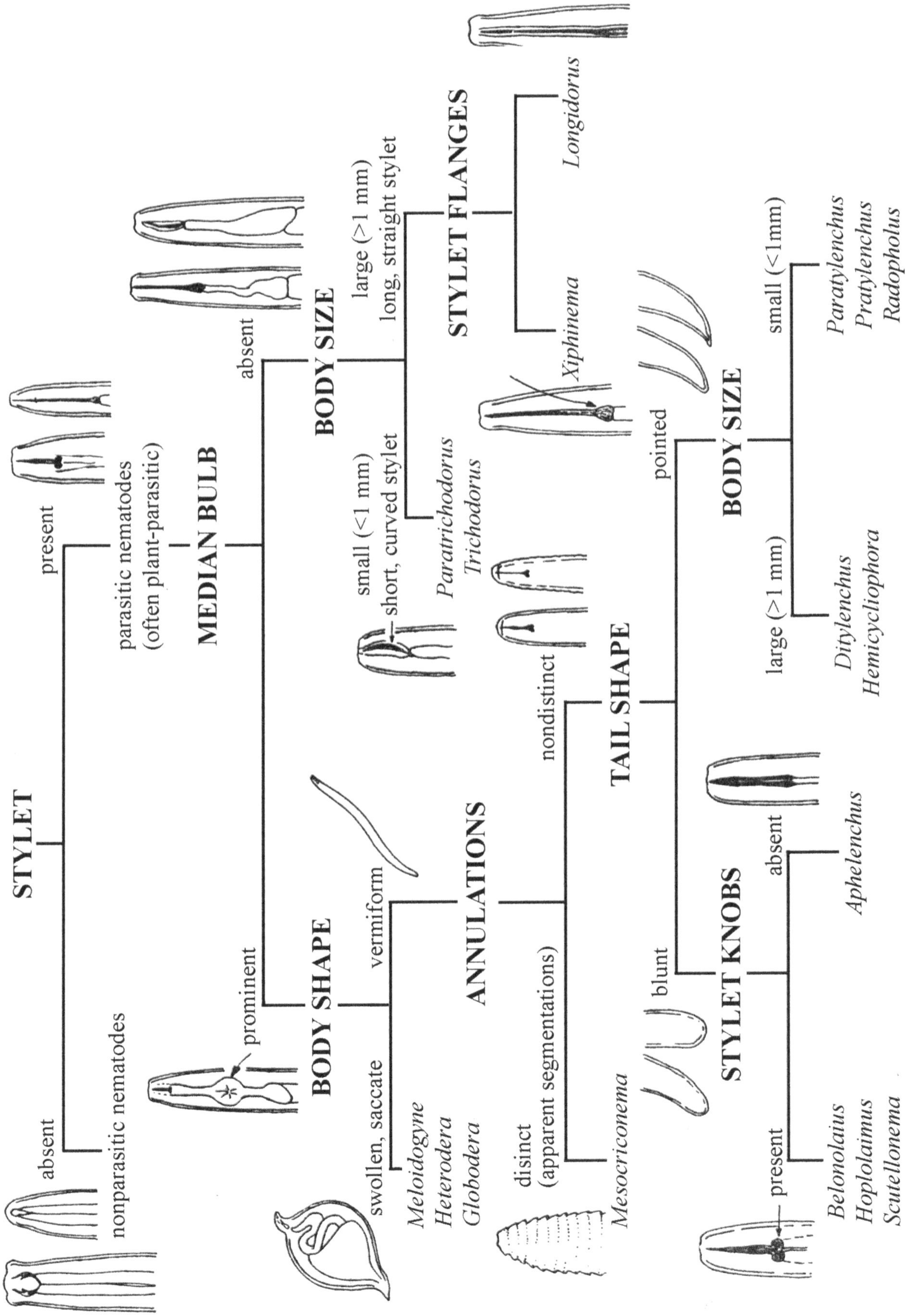

Fig. 23.1 A key to a few groups of plant-parasitic nematodes (after Baudoin).

Investigation 24

Optimal Use of the Light Microscope

The light microscope is the most important instrument that is used to study the morphlogy of nematodes. With its high quality oil immersion lense, it is very useful for examining nematodes to see all the minute details; however, proper use of the microscope is essential because the morphology of properly prepared specimens may be obscured or poorly resolved. Likewise, microscopes are expensive equipment and are easily damaged from improper use. If lenses are dirty or scratched from poor use, they may not provide an adequate image that is necessary to see all of the minor details of nematode morphology.

Objectives of the Laboratory

Upon completion of this lab the student should be able to:

1. **Adjust** a light microscope with Köhler illumination.
2. **Calibrate** an ocular microscope.
3. **Place** a glass slide with a mounted nematode under oil immersion in the light microscope.
4. **Clean** oil off an oil immesion lens.

Materials needed for this laboratory:

Nematode species:

Various genera and species

For each student:

Light microscope
Slide storage box
Nematode pick

For each pair of students:

- Alcohol lamp
- Butyl acetate
- Cheesecloth
- 2% formalin
- Glass coverslips
- Glass microscope slides
- Immersion oil
- Matches
- Ocular micrometer
- Nail polish
- Slide labels
- Slide ringer
- Small paint brush
- Stage micrometer
- Syracuse dishes
- Water

Work Sheet

I. Adjust your microscope for Köhler illumination.

A. Focus the microscope on a contrasting slide (plant sections stained with safranin and fast green are excellent) with a low power objective (about 10x).

B. Partially close the substage diaphragm and turn down the substage until the edges of the diaphragm are in focus; center with adjusting screws on the side of the substage. (Some microscopes have "pre-centered" con densers that should only be adjusted by a service person.) Rack up the substage condenser close to the stage. Close the field diaphragm (in front of the illumination) and move the substage condenser up or down until the edge of the field diaphragm is in perfect focus. The stained slide should remain in focus. Center the field diaphragm image by moving the mirror as necessary. (This adjustment can probably be omitted with a built-in illumination. However, some such microscopes do allow for minor adjustments of a mirror). Check alignment of the lamp relative to the microscope. When the field diaphragm is centered, open to just the edge of the field of view.

C. Open the substage diaphragm to the proper degree. This can be done in a very clean microscope room by removing the eyepiece and opening the diaphragm so that the back (upper) lens of the objective is just filled with light. A second method does not involve removing the eyepiece. Simply open the diaphragm to the exact point at which further opening does not increase the brightening of the field. When locating low contrast material (i.e., unstained nematodes), it may be necessary to further close the substage diaphragm to increase contrast. The microscopist should be aware that contrast achieved in this way reduces resolution. It is poor microscopy to use the position of the substage or condenser diaphragm or the field diaphragm to regulate brightness. Filters or an illuminator reostat is needed for this purpose.

D. Minor alignment adjustments will need to be made with objectives of higher magnifications. Certainly the field and condenser diaphragms will need to be reduced to fit the field, and slight focusing and centering adjustments may be needed. At very low magnifications, it may be necessary to swing out a condenser front lens or otherwise disperse light to fill the field.

II. Calibration of an ocular micrometer.

The ocular micrometer is used to measure the size of microscopic specimens. Because we use four different lenses, each with a different power of magnification, and because the optics of each microscope are slightly different, the ocular micrometer must be calibrated for each objective and each microscope. This standardization procedure is accomplished tbrough the use of a precision stage micrometer.

1. Record the serial number of your ocular micrometer. Be sure to always use the same micrometer throughout the term.

2. Set up your compound microscope with proper illumination.

3. Focus on the slide micrometer (white scale of known length).

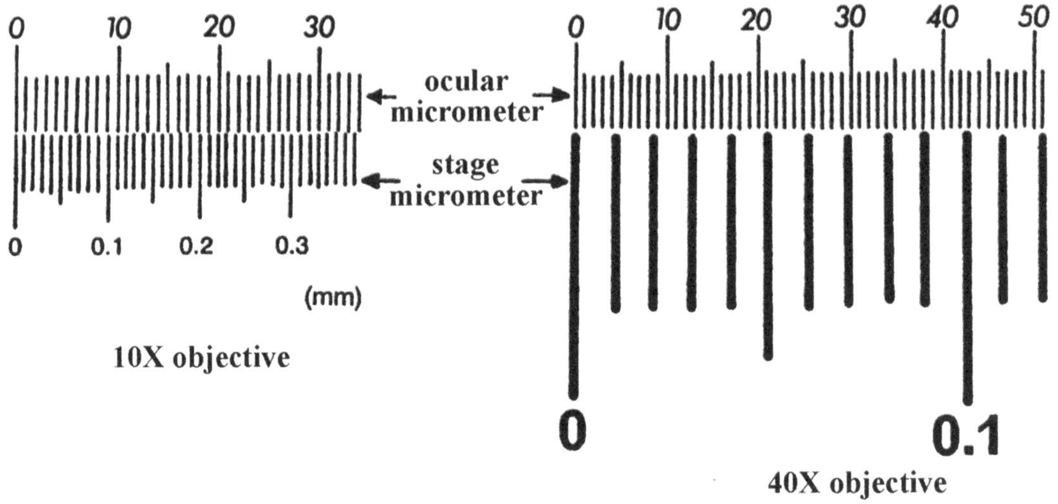

Fig. 24.1. Ocular micrometer calibrated with a stage micrometer. With the 10X objective, the tenth line on the ocular micrometer is in line with the eleventh line on the stage micrometer. Since the lines on the stage micrometer are 10 μm apart, (111/10) X 10 = 11 μm per ocular scale division.

4. Remove right eyepiece and insert an ocular micrometer. Handle both the eyepiece and ocular micrometer with care, and store in a dustfree environment.

5. Refocus on the white scale on the slide micrometer.

6. Determine the length of the black scale on the ocular micrometer by measuring it with the known units on the stage micrometer. Determine the length of the subunits on the ocular micrometer scale.

7. Calculate and record the following:

   a. Total length of scale on eyepiece micrometer.
   b. Length of 10 scale increments on eyepiece micrometer.
   c. Length of individual scale increments on eyepiece micrometer.

8. Repeat procedure for low power, high power, and oil immersion objectives.

9. Record the following data:

Microscope number: ______________ Ocular micrometer number: ______________

	Calibration data (1 mm = 1,000 μm)		
	Length of entire scale	Length of large units	Length of small units
Scanning lens	________	________	________
Low power lens	________	________	________
High dry lens	________	________	________
Oil immersion lens	________	________	________

10. Measure the following for an unknown nematode specimen:

   a. Nematode length ______________

   b. Nematode width ______________

   c. Stylet length ______________

   d. Nematode width ______________

III. Using an oil immersion lens.

A. Oil immersion lenses are absolutely necessary for the observation of nematodes.

1. Focus on the specimen with the high-dry lens.
2. Rotate the objective turrent half-way between the high-dry and the oil immersion lenses.
3. Place a small drop of oil on the top of the coverslip over the specimen.
4. Rotate the oil immersion lens into place.
5. Since the specimen will almost be in focus, use the fine focus not to gently bring the specimen into sharp focus.
6. Immediately after use, clean off the oil with a dry piece of lens tissue paper. At times it may be necessary to use a small amount of liquid lens cleaner.

IV. Complements to optimal use of the microscope.

A. The microscope, and specifically the lens, must be kept as clean as possible:

1. Remove dust with a camel hair brush.
2. Do not use incompatible immersion oils (some types do not mix).
3. Use lens paper with care to avoid scratches.
4. Do not use chemicals that leave a film on lens in the vicinity of microscopes.
5. Do not smoke near a microscope.
6. Do not use drinks or food where they might accidentally get on a microscope.
7. Mascara is a definite problem on eyepieces.
8. A very dirty microscope, particularly one with dirt inside, should only be cleaned by a service person.
9. Some solvents may actually dissolve the material seating the lens.

B. Eyeglasses normally should not be used with the microscope. However, nearsighted persons will need corrective glasses for camera lucida work. Persons with astigmatism should also wear glasses. High eyepoint eyepieces are available for microscopes of persons who must wear eyeglasses.

C. Microscope safety

1. Always carry a microscope with both hands.
2. Always return the turrrent to low power before removing the specimen.
3. Use only lenspaper with lens cleaning solution on the oil immersion lens.
4. Avoid getting oil on lenses other than the oil immersion lens.
5. Clean oil immersion lenses with lenspaper after each use.

Investigation 25

Preparation of Nematodes for SEM

Scanning electron microscopy (SEM) is very useful for examining nematodes to see all the minute details of their surface morphology. Proper preparation is essential because the morphology of poorly prepared specimens may be obscured by surface precipitation and distorted by shrinkage or swelling.

Preparing nematodes for SEM involves killing, fixation, dehydration, drying, mounting, coating, and viewing. All of these steps can be deleterious to the surface morphology by causing the nematodes to shrink and to produce other artifacts. The best way to insure that the artifacts are minimal is always to compare the SEM observations with images seen with a light microscope.

Objectives of the Laboratory

Upon completion of this lab the student should be able to:

1. **Prepare** a nematode processing chamber for preparing nematodes for SEM.
2. **Process** nematodes for SEM.
3. **Mount** nematodes for observation with the SEM.
4. **Observe** nematodes with the SEM.

Materials needed for this laboratory:

Nematode species:

Various genera and species

For each student:

Dissecting microscope
Nematode pick

For each pair of students:

BPI dishes
Double-sided cellophane tape
Epoxy embedding capsules (BEEM®)
10% formalin in buffer
Glass Stendor dish
4% glutaraldyhyde in buffer
Hole punch
Nylon mesh screen
Pipettes
Plastic Petri dish
SEM mounting stubs
Single-edged razor blade
Small round glass coverslips
Suitable buffer

8-step graded series of ethanol of 5, 10, 20, 25, 50, 75, 90, 95, and 100%

Work Sheet

I. Make a processing chamber for preparing specimens for SEM.

A. Cut off the conical end of a BEEM® capsule.

B. Punch a small hole in the cap.

C. Cut out a small square of nylon mesh screen and place it over one end of the capsule.

D. Place the cap into position over the screen to hold it in place.

1. What is the purpose of the processing chamber for preparing nematodes for SEM? __

____

____

____

____

____

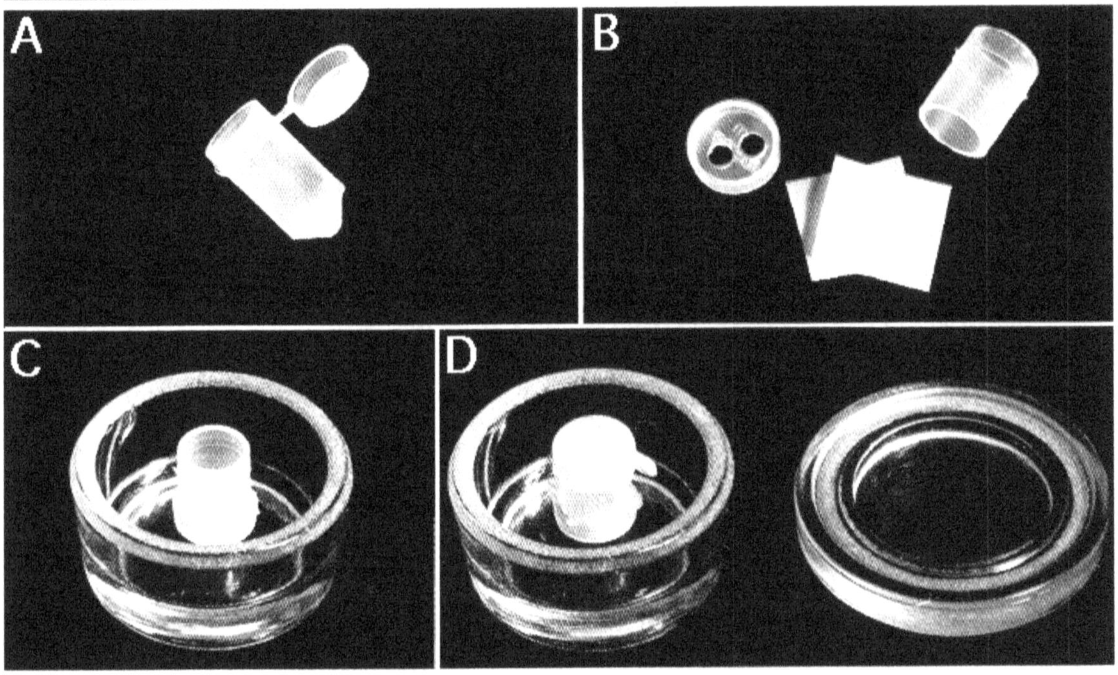

Fig. 25.1 SEM processing chamber for handling nematodes. A. BEEM® capsule container B. Capsule with conical portion removed and perforated cap and two squares of nylon mesh screen C. Chamber without cap in a Stendor dish D. Container with cap and lid of Stendor dish (After Eisenback)

II. Prepare nematodes for SEM.

A. Put 10 drops of clean tap water into a BPI dish.

B. Transfer the live nematode specimens into the water.

C. Chill the specimens in a refrigerator for 25 min at 8°C.

D. Place one drop of a mixture of 4% glutaraldehyde/10% formalin onto the specimens every 15 min until 10 drops of fixative have been added.

E. Allow the specimens to fix for at least 4 hr, overnite if possible.

F. Rinse the specimens in buffer three times for 15 min each.

G. Post-fix the specimens in 2% osmium tetroxide for at least 4 hr, overnite if desired.

2. What is sequential fixation? ____

____

3. *What is the philosophy for using sequential fixation?* ______________________

____

____

H. Dehydrate nematodes in a eight-step graded series of ethanol of 5, 10, 20, 25, 50, 75, 90, 95, and 100% ethanol for 15-20 min each.

I. Exchange the 100% ethanol three times to ensure that all of the water is removed.

J. Dry the specimens with a critical point dryer using proper techniques as descibed by the manufacterer of the drying apparatus. (Arrange a time for using the critical point dryer with the instructor.)

K. Store the dry nematodes in a dessicator.

4. *Why is it necessary to remove all of the water from the nematodes for SEM?*

____

____

____

5. *Why does air drying from water cause so much damage to the specimens?* __________

____

____

III. Mount the nematodes on stubs for viewing with the SEM.

A. Stick a small square of double sticky tape onto a SEM veiwing stub.

B. Attach a short length of hair across the tape on the stub.

C. Place nematodes on the tape propped up against the hair. The specimens are best viewed at a 90°angle with the surface of the stub.

D. Coat the specimens with a 20nm layer of gold/palladium. (Arrange a time for using the gold coater with the instructor.)

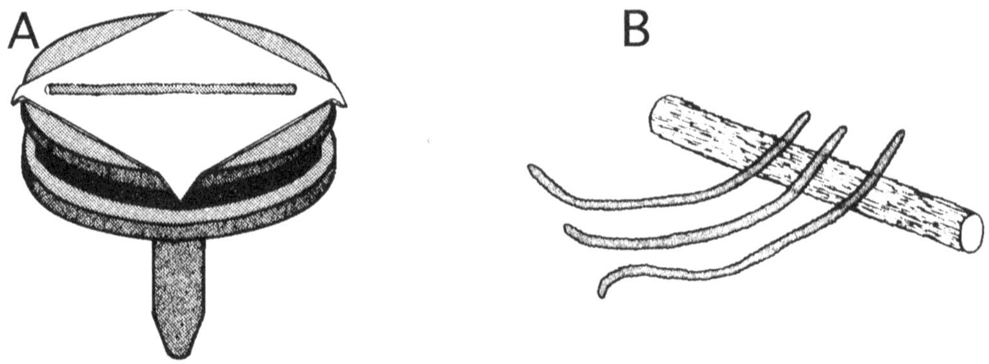

Fig. 25.2. Mounting whole dry specimens on the SEM viewing stub. A. Stub with square of double sticky tape and short length of human hair B. Closeup of nematodes propped up against the hair

6. Why are the specimens coated with a thin layer of gold/palladium? ______________

____

____

IV. Observe the specimens with the SEM.

(Arrange a time for using the SEM with the instructor.)

7. Compare and constrast the images of nematodes formed with a light microscope and a scanning electron microscope. ____

____

____

____

____

____

Appendix 1

Glossary of Nematological Terms

"a" - body length/greatest body width (range).

acrosome - refringent body in sperm of some animal parasites.

alae - expansions or projections formed by a longitudinal thickening of the cuticle; there are three types: 1) cervical alae - confined to the anterior region of the nematodes parasitic on animals; 2) caudal alae - occur in the posterior region of males in a number of genera; 3) longitudinal alae - usually 4 in number and extend the length of the body sublaterally.

allotype - a paratype of the opposite sex to the holotype.

amalgamate - uniting to form a unit.

ambulatory setae - bristles or setae, sometimes hollow and tube-like, by which some nematodes "walk" along a surface.

amphid - a chemo-sensory organ, occurring laterally in pairs, located in the anterior region of the nematodes; among the Secernentea, the amphids appear as small, pore-like openings on the lips; the openings are post-labial in the Adenophorea, and may be of varied form and made up of the following structures:

1. **amphidial duct** - a channel which connects the amphidial pore to the sensilla pouch.
2. **amphidial gland** - a microvillous organ enveloping sensilla pouch.
3. **amphidial nerve** - a nerve innervating the amphid.
4. **amphidial pore** - the aperture in the cuticle through which the amphid opens exteriorly.
5. **amphidial pouch** - (amphidial chamber) - a chamber, usually just behind amphidial pore, outline often cup-shaped.
6. **sensilla pouch** - chamber containing nerve processes.

amphidelphic (ovary) - having two ovaries, directed anteriad and posteriad, respectively.

ampulla - widening in canal, forming a reservoir.

aphelenchoid - dorsal gland empties the lumen of esophagus just anterior of the valve of the median bulb instead of opening near stylet base.

annule - the thickened interval between transverse depression (annulations) in the cuticle.

arcuate - curved like a bow.

areolation - a condition where the transverse body annulation traverses the lateral field.

articulate - jointed, segmented.

attenuate - to thin in consistency.

axenic culture - growth and reproduction of a single species on a nonliving medium.
axial - belonging to, or around, in the direction of, or along an axis.
"b" - body length/distance from anterior end to junction of esophagus and intestine (range).
bacillary layer - (now microvilli) the internal surface covering of the intestinal cells which is composed of microvilli that increase surface area.
basal region of esophagus - posterior portion of esophagus which may be variable in shape.
bifid - divided into two equal lobes or parts by a medium cleft.
bifurcate - divided into two equal lobes or parts by a medium cleft.
bilateral symmetry - that condition whereby the two halves of the body are exact counter-parts, one of the other (mirror images if divided in one plane); bilaterally symmetrical organisms are characterized by a ventral and dorsal side which are marked by openings.
blastocoel - the cavity of the blastula.
body pores - series of minute, slightly submedial or lateral depressions (or openings) traversing each side of the body (superfamily Dorylaimoidea).
bristle - setae.
buccal capsule - the structure connecting the oral opening with the anterior portion of the esophagus; the buccal capsule or stoma, as it is sometimes called, is subject to great variation among different nemas; the shape may be cylindrical or sub-globular; it may be toothed or armed with a protrusible spear or stylet.
bulboid - see esophagus.
bursa - the caudal alae of some males used to clasp the female during copulation.
"c" - body length/tail length.
caecum - a cavity open at one end, as the blind end of a lumen or duct (refers mostly to an outpocketing of the intestine at the anterior end).
cap cell - apical cell of female and male gonad.
cardia - a valvular apparatus concerning the esophagus and intestine (sometimes called the cardiac valve or esophago-intestinal valve).
caudal - pertaining to or located near the posterior region or tail.
caudal alae (bursa) - alae confined to the posterior part of the male nematode body.
caudal glands - 3-5 hypodermal glands located in the tail and emptying subterminally or terminally through a pore, the spinneret; these glands are known to occur in some of the Adenophorea; they secrete a fluid which hardens to form an attachment tread in water.
caudalid - a narrow, clear, bandlike area below the cuticle, located slightly anteriad of theanus in some Tylenchida (similar to hemizonid; may correspond to the paired ano-lumbar commissures linking the preanal ganglion to the lumbar ganglia).
cephalic - pertaining to or located near the head.
cephalic framework - the sclerotized framework of the head.
cephalids - two structures (posterior and anterior) situated in the cephalic region and extending in a complete circle around the body, possible part of nervous system.

chord - longitudinal internal thickening of the hypodermis (dorsal, 2 lateral, ventral).
circomyarian - cylindrical type of muscle cell in which the contractile region is perpendicular to the cell membrane all around the cell, enclosing a central core of sacroplasm.
cirri - elaborate cephalic appendages in front of the setae.
clavate - club-shaped.
cloaca - a common duct formed by the junction of the digestive and reproductive systems in males.
coelomocytes - cells lying in the pseudocoelom, apparently of mesenchymatous origin (e.g., anthrocytes, phagocytes, stellate cells, pseudocoelomocytes).
coelomyarian - type of somatic muscle cell in which the contractile region extends over the base and sides of the cell (this muscle cell is taller than wide).
commissure (latero-ventral, rectal) - connecting bands of nerve tissue.
concentered - radial muscle fibers attached to esophageal radii in certain points.
connate - congenitally or firmly united.
corpus - the anterior cylindrical part of the esophagus; the basal portion of the corpus at times may be swollen to form the metacorpus.
crustaformeria - glandular region of distal part of uterus, may play a role in the formation of egg envelope.
cuirasse - helmet-like appearance of the cuticle of the head.
cuticle - (inner, middle and outer layers) the non-cellular exterior covering of nematodes.
cyathiform (amphid) - pocket-like.
deirids - paired, pore-like organs located in the lateral fields in the vicinity of nerve ring of many of the Tylenchida; believed to be sensory in nature.
dentate - bearing teeth.
denticles - minute teeth or prickles.
depressor ani - the H-shaped muscle which serves to dilate the rectum and elevate the posterior lip of the anus, thus permitting defecation.
didelphic - possessing two complete genital tubes or ovaries.
dioecious - sexually distinct; having the male reproductive organs in one individual, the female in another.
diorchic - possessing two testes.
diplogasteroid - see esophagus.
dispersed - radial muscle fibers not attached to esophageal radii in certain points.
dorsal esophageal gland outlet (orifice) - the point at which the dorsal gland empties into the lumen of the esophagus.
dorylaimoid - see esophagus.
ecdysis - the act of shedding the outer cuticular layer together with internal cuticular linings; a molt.
ejaculatory duct - terminal portion of vas deferens that is provided with musculature.
ejaculatory glands - glands associated with the ejaculatory duct; secrete adhesive cement functional in copulation.

endoderm - inner "germ layer" that gives rise to intestine.
epithelium - a cellular, membrane-like tissue covering a free surface or lining of a cavity.
esophageal gland lobe - the basal, glandular portion (usually 3 glands) of many plant-parasitic nematodes; it may overlap the intestine.
esophago-intestinal valve - the valve between the esophagus and intestine.
esophagus (pharynx) - the portion of the alimentary canal between the buccal capsule or stoma and the anterior portion of the intestine; types of esophagi are as follows:

1. **aphelenchoid** - dorsal gland empties the lumen of esophagus just anterior of the valve of the median bulb instead of opening near stylet base. Example - *Aphelenchoides*.
2. **bulboid** - long, slender anterior portion with base forming a distinct, muscular bulb. Example - *Achromadora*. (Plectoid similar but basal bulb has true valve. Example - *Plectus*.)
3. **cylindrical** - esophagus proper is approximately the same diameter at stoma all way back to intestine. Example - *Monochus*.
4. **diplogasteroid** - anterior portion has a muscular bulb with posterior portion forming a glandular bulb. Example - *Diplogaster*.
5. **dorylaimoid** - long, slender, anterior portion and thicker, long, bulbar portion muscular and glandular. Example - *Dorylaimus*.
6. **rhabditoid** - true basal bulb with valve with anterior portion having a swelling or pseudobulb. Example - *Rhabditis*.
7. **tylenchoid** - a dorsal gland opens near base of stylet, not over one-third of distance back of stylet base; median bulb with basal glands that may (*Tylenchorhynchus*) or may not form a distinct bulb (*Pratylenchus*).

esophagus lumen - the canal of the esophagus.
excretory canal - a tube-like structure located in the lateral chord or pseudo coelom that collects excretory products and conveys them to excretory duct and pore in turn.
excretory duct (termonal duct) - a tube or canal, lined with cuticle, that leads to the excretory pore.
excretory pore - the exterior opening of the excretory system; it is generally located on the ventral side of the body near the base of the esophagus. (Also known as the orifice of the cervical gland.)
excretroy sinus - a common cavity of the excretory canals located ventrally and communicating to the exterior by the excretory pore via a cuticularly lined reservoir or an elongated terminal duct.
external cortex - the outer layer of the cuticle.
fiber layer - the inner layer of the cuticle.
filiform - having the shape of a thread.
flexure - a turn or fold.
fossores - modified structures of head, dorsal and ventral lips, fitted for "digging." Example - *Diploscapter coronata*.
furcate - branching like a fork.

fusiform - spindle-shaped; tapering at each end.
ganglion (dorsal, lateral, amphidial, retroversicular, etc.) - a group of nerve cell bodies, with their conspicuous nuclei.
genital papillae - sensory organs of a tactile nature locate on the male tail.
genital primordium - the initial cells of the reproductive system.
germinal zone of ovary or testis - a region in which rapid division of the germ cells takes place.
glottoid apparatus - a sclerotized, valvular structure at the base of the stoma.
growth zone of ovary or testis - a region of gradual increase in size of the oogonia or spermatozoa.
gubernaculum - spicule guide, sclerotized accessory piece.
guiding ring - a sleeve-like structure which surrounds and guides the stylet in certain genera of the Dorylaimoidea; position varies among the genera from apex to posterior portion of the stylet. (End of stomatal cavity where cuticle living is attached to stylet.)
hemizonid - a lens-like structure situated between the cuticle and hypodermal layer on the ventral side of the body just anterior to the excretory pore; generally believed to be associated with the nervous system (commissure).
hemizonion - companion structure to hemizonid (smaller and posterior to hemizonid).
hermaphrodite - an individual having both functional male and female reproductive organs.
hirsute (body) - rough with hair or bristles.
hologonic - the germ cells are formed along the entire length of the gonad.
holomyarian - an arrangement of the somatic musculature in nematodes with only 2 chords or no chords, where the musculature is divided into 2 fields or is continuous.
holotype - the single specimen designated or indicated as "the type" by the original author at the time of the publication of the original description.
hypodermal glands - glands formed by the hypodermis (caudal glands, lateral hypodermal glands, amphidial and phasmidial glands).
hypodermis (epidermis in some) - a thin tissue layer beneath the cuticle which thickens to form the dorsal, lateral and ventral chords which extend the length of the body (also forms the cuticle).
incisures - the longitudinal markings within the lateral fields (sometimes called involutions; also called lines).
innervation process - a process of the nematode muscle cell that extends to the motor nerve (neuromuscular process).
intersex - an individual more or less intermediate in phenotype between male and female (shows secondary male or female characters).
intestino-rectal valve - the sphincter muscle separating the intestine from the rectum or cloaca.
isthmus - relatively narrow portion of esophagus just anterior to the basal region.
"L" - mean body length in mm (range).
labial - pertaining to the lips.

lateral field - longitudinal cuticular thickening situated on top of the lateral chords; the field may be divided by longitudinal striae (= incisures) and at times by transverse markings (areolation).

lateral guiding pieces - structure that consists of two, very small, lineate, cuticularized pieces lying lateral to the distal portion of the spicules and joining with the muscular sheath surrounding the anus (present in Dorylaimidae, Belondiridae and Leptonchidae).

lectotype - one of a species of syntypes which, subsequent to the publication of the originial description, is selected and designated through publication to serve as the type.

leptoderan - caudal alae which do not meet posterior to the tail tip.

lip - cuticular structures (usually six), 2 subdorsal, 2 lateral, 2 subventral, surrounding the mount opening. (May be fused in pairs.)

longitudinal ridges - raised cuticular areas which extend through the length of the body and are present on all sides of the nematode.

lumen - the canal or duct of the esophagus.

mammillate - digitate; nipple-shaped protuberances.

marginal muscle fibers - muscle fibers of the esophagus at the apices of the triradiate lumen which have a suspensory and skeletal function.

matrix - the middle layer of the cuticle.

median bulb - see metacorpus.

meromyarian - an arrangement of the somatic musculature, where only few (2-4) longitudinal rows of muscle cells are present between each two chords.

metacorpus - the swollen posterior portion of the corpus of the esophagus (sometimes referred to as the median bulb).

metarhabdions - see rhabdions.

metatype - a specimen compared by the author of the species with the holotype and determined by him as conspecific with it.

micropyle - the minute opening in the membrane of an egg through which the spermatozoon enters.

molt - to cast off the cuticle.

monodelphic - possessing one genital tube or ovary.

monorchic - possessing one testis.

monoxenic culture - growing an organism in the presence of another organism.

mucro - a stiff or sharp point abruptly terminating an organ.

mucronate - ending in a sharp point.

mural - resembling a wall.

myofilaments - the thick and thin muscle elements of the contractile region of the muscle cell.

myriocytous - a very large number of cells (over 8,224) in the intestinal epithelium (the cells are cuboidal).

neck - that portion of the nema body occupied by the esophagus.

neotype - a specimen selected as type subsequent to the original description in cases where the primary types are definitely known to be destroyed.

nerve ring - a circum-esophageal commissure, or the center of the nervous system of nemas which encircles the esophagus; it is composed largely of nerve fibers and associated ganglia.

ocelli - eyelike spots (pigment spots).

oligocytous - a small number of cells (up to 128) in the intestinal epithelium (the cells are rectangular).

onchiostyle (odontostylet) - see stylet.

onchium - pharyngeal tooth.

opercula - less sclerotized areas at the poles of hard-shelled eggs of some animal parasites allowing the escape of the larva.

opisthodelphic - having the uterus (uteri) directed posteriad.

ovejector (ovijector) - the elongated, heavily muscularized vagina in some animal parasites, aids in the expulsion of eggs.

oviduct - a short usually tubular, thickwalled, part of the female reproductive system between ovary and spermatotheca.

oviparous - producing eggs that hatch after exclusion from the body.

ovoviviparous - production of eggs which hatch in the mother's body.

papillae - tactile, sensory organs found on various body regions (labial, cephalic, etc.)

paratype - a specimen other than the holotype which was before the author at the time of preparation of the original description and was designated or indicated as such.

parathenogenesis - reproduction by the development of an unfertilized egg.

peloderan - (bursa) caudal alae which meet posterior to the tail tip.

perineal pattern - cuticular pattern in perineal area (area of vulva and anus) of *Meloidogyne* females; composed of fine, wavy, fingerprint-like striations that are more or less species characteristic.

pharynx (esophagus) - the muscular feeding tube that connects the mouth with the intestine.

Phasmid - pore-like structure located in the lateral fields of the posterior region of nematodes belonging to the class Secernentea: function is believed to be sensory (sometimes called precaudal glands).

platymyrian - type of somatic muscle cells in which the contractile region is limited to the side of the hypodermis (this muscle cell is wider than tall).

polar filament - a projection of the protein membrane on some eggs of animal parasites, probably serving in attachment.

polycytous - a large number of cells (128-8,224) in the intestinal epithelium (the cells are hexagonal).

polymyarian - an arrangement of the somatic musculature, where a large number of longitudinal rows of muscle cells is present between each two chords.

polynucleate - having more than one nucleus per cell.

postvulval-uterine sac - a rudimentary or vestigial ovary which may serve to store spermatozoa (often called postvulvar uterine sac).

pre-rectum - a differentiated portion of the posterior end of the intestine of the Dorylaimidae, Belondiridae and Leptonchidae; it is separated from the main intestine by a constriction of the lumen; the granules of its cells are smaller, less numerous and much lighter in color than those of the intestine.

probolae - prominent, sometimes ornate, structures on the head; often modifications of the labial and cephalic regions.

procorpus - the cylindrical portion of the corpus anterior to the metacorpus (of the esophagus).

prodelphic - having the uterus (uteri) directed anteriad.

prorhabdions - see rhabdions.

protandry - the gonad first produces sperms that are stored and later fertilizes the eggs subsequently developed by the same gonad.

pseudocoelom - the body cavity of a nematode which is not lined by epithelium (mesoderm) and in which the various internal organs are suspended in a fluid; the body fluid apparently functions as a respiratory and circulatory system.

punctation - minute dots or ovals in deeper layers of the cuticle which occur in transverse or longitudinal rows and may be arranged in various patterns.

pyriform - pear-shaped.

rachis - a central strand of non-nucleated tissue to which oogonia are attached in the ovary (function uncertain).

radial muscle fibers - muscle fibers of the esophagus which act to dilate the lumen of the esophagus.

rays - muscular genital papillae or tactile organs located within the bursa (primarily in animal parasites).

recurved - to curve in an opposite direction or bend back.

rectal glands - three to six large glands opening into rectum of some Secrenentea; function in case of *Meloidogyne* spp. is the secretion of the gelatinous matrix of the egg-mass.

rectum - posterior gut of female; a narrow, dorso-ventrally flattened tube that is lined with cuticle and separated from the intestine by a sphincter muscle.

reflexed - bent back.

renette - a cell or group of cells emptying through the excretory duct(s) in Adenophorea.

reniform - kidney-shaped.

reticulate - having cross markings like a net.

retrorse - bent backward or downward.

revolute - rolled backward or downward.

rhabdions - the plates of the cuticular lining of the stoma which make up the walls of the various divisions of the stoma in free-living Rhabditida as follows:

1. **cheilorhabdions** - most anterior parts of the stomatal wall and correspond to the entrance to the stoma.
2. **protorhabdions** - the walls of the cylindrical stoma; which may be subdivided as follows: Prorhabdions; Mesorhabdions; and Metarhabdions.
3. **telorhabdions** - that portion of the stomatal wall (Rhabditis) which serves as a connecting valve between stoma and esophagus.

rhabditoid - see esophagus.
rugae - full of wrinkles.
"S" - stylet length/body diameter measured at the base of the stylet.
saccate - pouch or sac-like.
sclerotized - hardened cuticular (refractive) regions.
scutellum - a large phasmid. Example - *Scutellonema.*
seminal receptacle - the organ in which sperms are stored in the female reproductive system.
sensilla - the group of nerve endings in the sensilla pouch.
setae - elongated cuticular structures articulating with the cuticle; in general, tactile sensory organs usually located around the oral openings (such as labial, somatic or caudal).
sexual dimorphism - pronounced difference in the morphological expression (form) of the two sexes of a single species.
somatic musculature - consists of a single layer of more or less spindle-shaped cells which are attached to the hypodermis throughout their length.
spear - synonymous with stylet.
spermatheca - the organ in which sperms are stored in females.
spicate - having the form of a spike (tail).
spicules - paired, sclerotized structures which are the male copulatory organs.
spination - having spines on the body surface; example - *Criconema.*
spicular pouch - a pouch, containing the spicules, which is often called the sheath in parasitic nematodes.
spinneret - a pore-like opening of the caudal glands situated in the tail, located terminally or subterminally in some nemas belonging to the Adeonophorea.
spiral - referring to the coil-like type of amphidial opening.
squamous - covered with or consisting of scale.
stoma - the portion of the digestive tract between the oral opening and esophagus.
striae - superficial markings of the cuticle, appearing as grooves or clefts; if present, may encircle lips or body (longitudinal and transverse).
striation - fine transverse markings of the cuticle.
stylet - hollow protrusible spear used to puncture plant or animal prey:

1. **stomatostyle** - a spear which apparently has developed gradually through evolution by the fusion of sclerotization of the stomatal walls; commonly found among many species of Secernentea.
2. **odontostyle** - this type of spear represents a large tooth which originates in the esophageal wall, generally found among some members of the Adenophorea.

stylet aperture - the anterior opening of the stylet; located dorsally in the Dorylaimoidae and generally ventrally in the Tylenchida.
stylet extension (odontostyle extension or odontophore) - the sclerotized structure extending between the esophagus and the dorylaimoid spear.

stylet knobs - the basal protuberances (usually three) of the spear.
subsidiary - secondary, additional.
supplements (supplementary organs) - genital papillae, located preanal on the ventral side of the male; derived from cuticle, but may be provided with glands, and functioning during copulation.
syncytium - a multinucleate cell.
syntype - one of a number of specimens of equal nomenclatural rank which formed all or part of the material before the original author, in those cases where the author did not designate or indicate a holotype.
"T" - percent of body occupied by male reproductive system which is expressed: T = 65. (Length of male reproductive system expressed as a percentage of total body length.)
tail - the portion of the body between the anus and the posterior terminus.
telogonic - the germ cells are profilerated in the proximal end of the gonad.
telorhabdions - see rhabdions.
tessellate - checkered; a type of cuticle pattern in which the longitudinal ridges are broken by transverse striations into transverse rows of squares.
topotype - a specimen collected at the type locality.
trifurcate - bearing three branches or forks.
triploblastic - possessing three germ layers: ectoderm, mesoderm and endoderm.
triradiate - having three radiating arms or branches.
truncate - having the end square or even.
tylenchoid - see esophagus.
type - a zoological object which serves as the base for the name of a taxonomic category (e.g., a specimen which is the name bearer for a species, a species which is the name bearer for a genus, etc.).
uninucleate - having one nucleus per cell.
"V" - position of the vulva from the head expressed as a percentage of the total body length.
vagina - a canal, lined with cuticle, that connects the uterus (uteri) with the female gonopore.
valve - a structure which regulates the rate and/or direction of intake of materials; in nemas, an example is the esophago-intestinal valve or cardia.
vas deferens - a slender tube-like male gono-duct; unites posteriorly with the rectum to form a cloaca (may be muscular or glandular or both).
vas efferens - distal, specialized tubular part of the seminal vesicle connecting growth zone of testis and seminal vesicle.
vestibule - a chamber of passage.
vestigial - pertaining to structures or organs that were well developed in an animal's ancestors but have become rudimentary during the course of evolution.
viviparous - bearing living young.
vulva - the exterior opening of the female's reproduction system (female gonopore) which appears as a transverse slit on the ventral side of nematodes.

warts - development of cuticular crests that may be arranged in single or double longitudinal rows on one side of the body; occur on some free-living nemas such as Bunonema.

wing - longitudinal structure in cuticle, or projecting from it, usually lateral, but sometimes numerous and evenly spaced around the nema.

(prepared by K. R. Barker and H. Hischmann)

Appendix 2

Nematode Chart

Scientific Name	Common Name	Principal Hosts
	CYST FORMING NEMATODES Nematodes that form cysts covering the eggs	
HETERODERA	LEMON-SHAPED CYST NEMATODES	
H. avenae	Oat nematode	Small grains, corn, grasses
H. carotae	Carrot root nematode	Carrot
H. cruciferae	Cabbage root nematode	Cabbage, other crucifers
H. glycines	Soybean cyst nematode	Soybean
H. goettingiana	Pea root nematode	Pea
H. humuli	Hop nematode	Hops, hemp
H. schachtii	Sugarbeet nematode	Cole crops, celery, cress, kale, weeds, radish, spinach, turnip, sugarbeet
H. trifolii	Clover root nematode	Clover, legumes
H. weissi	Knotweed nematode	Knotweed
H. zea	Corn cyst nematode	Corn
GLOBODERA	ROUND CYST NEMATODES	
G. rostochiensis	Golden nematode	Irish potato, tomato, eggplant
G. tabacum	Tobacco cyst nematode	Tobacco, tomato
G. pallida	Potato cyst nematode	Irish potato, tomato, eggplant
PUNCTODERA		
P. punctata	Grass cyst nematode	Wheat, other small grains
DITYLENCHUS	STEM AND BULB NEMATODES	
D. destructor	Potato rot nematode	Potato, dandelion
MELOIDOGYNE	ROOT KNOT NEMATODES	
M. arenaria	Peanut root knot nematode	1700 hosts, including
M. chitwoodi	Columbia root knot	potato, corn, alfalfa
M. exigua	Coffee root knot nematode	forage, small grains,
M. hapla	Northern root knot nematode	grasses, fruit, vegetables,
M. incognita	Southern root knot nematode	nursery weeds, field
M. javanica	Javanese root knot nematode	crops

MELOIDODERA CYSTOID NEMATODE

M. floridensis.......... Pine

NACOBBUS

N. batatiformis.......... Sugarbeet

PRATYLENCHUS ROOT-LESION OR MEADOW NEMATODES

P. brachyurus..........Smooth-headed nematode..........Potato, corn, peanut, cotton, pineapple, avocado, tobacco

P. coffeae..........Coffee nematode..........Coffee, tea, abaca, sugarcane, banana, olive, apple

P. minor..........Wheat, tobacco, corn

P. penetrans..........Nursery, grasses, forage, oats, strawberry, apple, tomato

P. pratensis..........Grains, grasses, strawberry, lily, narcissus

P. scribneri.......... Scribner's meadow nematode..........Potato, strawberry

P. thornei..........Thorne's meadow nematode..........Wheat, grains, grasses

P. vulnus..........Lesion nematode..........Ornamentals, trees, forage, fruit, beans, walnut, grape, olive, fig, boxwood, stone fruits, citrus, raspberry, rose

P. zeae..........Corn nematode..........Corn, wheat

RADOPHOLUS BURROWING NEMATODES

R. oryzae..........Rice nematode..........Rice

R. similis..........Burrowing nematode..........Citrus, ornamentals, banana

R. citrophilus..........Burrowing nematode..........Citrus, ornamentals, avocado

ROTYLENCHULUS RENIFORM NEMATODES

R. reniformis.......... Kidney-shaped or Reniform nematode.......... Turf, cotton, peanut, tomato, weeds, ornamentals

SPHAERONEMA

S. sasseri.......... Red spruce, Fraser fir

TYLENCHULUS

T. semipenetrans..........Citrus nematode..........Citrus, olive

ECTOPARASITIC NEMATODES

Nematodes that feed on the root surface and normally do not enter the root tissue

BELONOLAIMUS STING NEMATODES

B. gracilis ... Turf, strawberry, corn, ornamentals, cotton,

B. longicaudatus ... soybean, celery, vegetables, peanut, pine

CACOPAURUS

C. pestis ... Walnut nematode ... Walnut

CRICONEMA PINE NEMATODES Pine trees

MESOCRICONEMA RING NEMATODES

M. ornata ... Turf, cotton, peanut,

M. xenoplex ... citrus, weeds, deciduous fruit

DOLICHODORUS AWL NEMATODES

D. heterocephalus ... Celery, beans, corn, tomato, and water chestnut

HELICOTYLENCHUS TRUE SPIRAL NEMATODES

H. nannus ... Peanut, tobacco, tomato,

H. dihystera ... clover

HEMICYCLIOPHORA SHEATH NEMATODE

H. heterocephalus ... Conifers

HOPLOLAIMUS LANCE NEMATODES Turf, nursery, corn,

H. galeatus ... sugarcane, alfalfa,

H. columbus ... red clover, soybean

LONGIDORUS NEEDLE NEMATODES

L. elongatus ... Grasses, peppermint

PARATYLENCHUS PIN NEMATODES Celery, many vegetables,

P. projectus ... fig

ROTYLENCHUS SPIRAL NEMATODES

R. buxophilus ... Boxwood nematode ... Boxwood, ornamentals

PARATRICHODORUS	STUBBY ROOT NEMATODES	Turf, corn, celery,
P. christiei		cole crops, bean, pea, cotton, cowpea, beet, pepper, weeds, tomato
TRICHODORUS		(See above list)
TYLENCHORHYNCHUS	STYLET NEMATODES	
T. claytoni	Tobacco stunt nematode	Tobacco, corn, cotton, milo, strawberry, alfalfa, peanut, bean, ornamentals
XIPHINEMA	DAGGER NEMATODES	Cotton, strawberry, oats,
X. americanum		tobacco, peach, nursery
X. index		grape

ABOVE GROUND FEEDERS

Nematodes that feed within plant tissue above ground

ANGUINA	SEED GALL NEMATODES	
A. agrostis	Grass nematode	Grasses
A. tritici	Wheat nematode	Rye, wheat
APHELENCHOIDES	FOLIAR NEMATODES	
A. besseyi		Vanda orchid, causes white tip of rice, strawberry, African violet
A. fragariae	Bud and leaf nematode	Flax, mint, onions, oats, sweet potato, nursery, begonia, fern, iris, lily,
A. ritzemabosi	Bud and leaf nematode	strawberry, white clover, ribes, orchid, violet, chrysanthemum
BURSAPHELENCHUS		
B. xylophilus	Pinewood nematode	Pine trees
DITYLENCHUS		
D. dipsaci	Stem and Bulb, or Teasel nematode	Alfalfa, clover, onion, garlic, sweet potato, strawberry, nursery

Appendix 3

Nematological Journals

*Afro-Asian Journal of Nematology (1990-1996)
Annals of Applied Biology
Annals of the Phytopathological Society of Japan
Biological and Cultural Tests for Control of Plant Diseases
Bulletin OEPP = EPPO Bulletin
Canadian Journal of Zoology
Comparative Parasitology (2000)
Fitopatologia
*Fundamental and Applied Nematology (until 1991 Revue de Nématologie)
Fungicide & Nematicide Tests
*Indian Journal of Nematology (1971)
International Journal of Nematology (until 1996 Afro-Asian Journal of Nematology)
*Japanese Journal of Nematology (1972)
*Journal of Nematology (1969)
*Journal of Nematode Morphology and Systematics (1999)
*Nemapix (1998)
*Nematologica (1956-1998)
*Nematologica Mediterranea (1973)
* Nematology: International Journal of Fundamental and Applied Nematology (1999)
*Nematropica (1971)
Netherlands Journal of Plant Pathology
*Pakistan Journal of Nematology (1983)
Parasitology
Phytoparasitica
Phytopathologica Mediterranea
Phytopathologische Zeitschrift (Journal of Phytopathology)
Phytopathology
Phytophylactica
Phytoprotection
Plant Disease (formerly Plant Disease Reporter)
Plant Pathology
Proceedings of the Helminthological Society of Washington (until 2000 = Comparative Parasitology
*Revue de Nematologie (1978-1991)
*Russian Journal of Nematology (1993)
Rivista di Patologia Vegetale
Systematic Parasitology

*specializes exclusively in plant nematology

TRANSLATIONS, ABSTRACT COMPILATIONS, AND OTHER SECONDARY SOURCES

Bibliography of Agriculture
Biological Abstracts
Bio Research Index
Bio Systematic Index
Commonwealth Agricultural Bureaux (CAB International) Abstract Journals
Current Contents (Life Sciences)
Dissertation Abstracts
Helminthological Abstracts, Series B
Nematological Abstracts (Helminthological Abstracts, Series B: Plant Nematology.)
Science Citation Index
Zoological Record

Appendix 4

Nematological Societies

PROFESSIONAL SOCIETIES FOR NEMATOLOGISTS

Afro/Asian Society of Nematologists
American Phytopathological Society
American Society of Parasitologists
Asociacion Peruana de Fitopatologia
Association of Applied Biologists
Brazilian Society of Nematologists
British Society of Parasitologists
Chinese Society of Nematologists
European Society of Nematologists
Japanese Nematological Society
Mediterranean Nematological Society
Nematological Society of India
Nematological Society of Southern Africa
Organization of Tropical American Nematologists
Pakistan Society of Nematologists
Russian Society of Nematologists
Sociedad Brasileira de Nematologia
Societa Italina Di Nematologia
Society of Nematologists
The Helminthological Society of Washington

Appendix 5

Nematological Resources

BOOKS

Anderson, R. C. 2000. Nematode Parasites of Vertebrates: Their Development and Transmission. 2nd ed. CABI Publishing: Wallingford. Oxon, United Kingdom.

Anderson, R. V., and R. H. Mulvey. 1979. Plant-parasitic Nematodes in Canada. Part 1. An Illustrated Key to the Genera. Monograph No. 20. Research Branch, Agriculture Canada: Ottawa, Canada.

Ayoub, S. M. 1980. Plant Nematology: An Agricultural Training Aid. Nema Aid Publications: Sacramento, California.

Barker, K. R., G. A. Pederson, and G. L. Windham. 1998. Plant Nematode Interactions. American Society for Agronomy, Crop Science Society of America, Soil Science Society of America: Madison, Wisconsin.

Barron, G. L. 1977. The Nematode-Destroying Fungi. Topics in Mycobiology. Canadian Biological Publications: Guelph, Ontario, Canada N1H 6J9.

Bilgrami, A L. 1998. Nematode Biopesticides. Biological Control of Nematodes. Aligarh Muslim University. Aligar 202002, India

Binns, M. R., J. P. Nyrop, and J. van der Werf. 2000. Sampling and Monitoring in Crop Protection: The Theoretical Basis for Designing Practical Decision Guides. CABI Publishing: Wallingford, Oxon, United Kingdom.

Bird, A. F., and J. Bird. 1991. The Structure of Nematodes. 2nd ed. Academic Press: New York.

Bongers, T. 1988. De Nematoden van Nederland. Publication Foundation Royal Dutch Natural History Society, Library No. 46. Hoogwoud: Nederlands.

Brown R. H., and B. R. Kerry. 1987. Principles and Practice of Nematode Control in Crops. Academic Press: London.

Burges, H. D., (ed.). 1998. Formulation of Microbial Biopesticides - Beneficial Microorganisms, Nematodes, Seed Treatments. Kluwer Academic Publications: Boston and Dordrecht.

Commowealth Institute of Helminthology 1972 - 1985. Descriptions of Plant-Parasitic Nematodes. Sets 1-8. 120 Species. Commonwealth Institute of Helminthology: St. Albans, United Kingdom.

Croll, N. A., and Matthews; B. E. 1977. Biology of Nematodes. Halsted Press.

Decker, H. 1997. Plant Nematodes and Their Control - Phytonematology. Brill Academic Publishers: Leiden.

Decreamer, W. 1995. The Family Trichodoridae: Stubby Root and Virus Vector Nematodes. Kluwer Academic Publications: Dordrecht.

Dindal, D. L. 1991. Soil Biology Guide. John Wiley & Sons: New York.

Dropkin, V. H. 1989. Introduction to Plant Nematology, 2nd ed. John Wiley & Sons: New York.

Eisenback, J. D. 2002. Identification Guides for the Most Common Genera of Plant-Parasitic Nematodes. Mactode Publications: Blacksburg, Viginia.

Evans, K., D. Trudgill, and J. M. Webster, (eds.). 1993. Plant Parasitic Nematodes in Temperate Agriculture. CAB International: Wallingford, United Kingdom.

Fenoll, C., F. Grundler, F. M. W. and S. A. Ohl, (eds.). 1997. Cellular and Molecular Aspects of Plant-Nematode Interactions. Kluwer Academic: Dordrecht.

Fortuner, R. 1988. Nematode Identification and Expert System Technology. NATO Workshop, Raleigh. Plenum Press: New York and London.

Fortuner, R., E. Geraert, M. Luc, A. R. Maggenti, and D. J. Raski. 1988. A Reappraisal of Tylenchina (Nemata). ORSTROM, France.

Gaugler, R., and H. K. Kaya. 1990. Entomopathogenic Nematodes in Biological Control. CRC Press: Boca Raton, Florida.

Gommers, F., and P. W. Th. Maas, (eds.). 1992. Nematology: from Molecule to Ecosystem. Proceedings of the Second International Nematology Congress 1990 Veldhoven. European Society of Nematologists.

Gubina, V. G. 1988. Nematodes of Plants and Soils: the Genus *Ditylenchus*. Saad Publications: Karachi, Pakistan.

Hickey, K. D. 1986. Methods for Evaluating Pesticides for Control of Plant Pathogens, Fungicides, Nematicides and Bactericides. American Phytopathological Society: St. Paul, Minnesota.

Hunt, D. J. 1994. Aphelenchida, Longidoridae and Trichodoridae: Their Systematics and Bionomics. CAB International: Wallingford, United Kingdom.

Jairajpuri, M. S., and W. Ahmad. 1992. Dorylaimida - Free-living, Predaceous and Plant-parasitic. E. J. Brill: Leiden.

Jepson, S. B. 1987. Identification of Root-knot Nematodes. CAB International: Wallingford, Oxon, United Kingdom.

Khan, M. W., (ed.). 1993. Nematode Interactions. Chapman & Hall: London, United Kingdom.

Khan, M. W., and Sharma, P. D., 1996. Root-knot Nematodes in India. Venus Publishing House: New Delhi, India.

Lamberti, F., and Taylor, C.E. (eds.). 1979. Root-knot Nematodes. Academic Press. New York.

Lamberti, F., C. de Giorgi, and D. McK. Bird. 1996. Advances in Molecular Plant Nematology. NATO ASI Series A., Vol. 268. Plenum Publishers: New York.

Lamberti, F. D. Bird, and C. de Giorgi. 1994. Advances in Molecular Plant Nematology. Kluwer Academic Publishers:

Lamberti, F., C. E. Taylor, and J. W. Seinhorst. 1975. Nematode Vectors of Plant Viruses. Plenum Press: London and New York.

Lorenzen. S. 1994. The Phylogenetic Systematics of Freeliving Nematodes. The Ray Society: Andover.

Loof, P. A. A. 1999. Nematoda, Adenophorea (Dorylaimida). *In*: J. Schwoerbel and P. Peter Zwick, (eds.). Begr. v. August Brauer. Süsswasserfauna von Mitteleuropa. 21 Bde. Urban & Fisher; Fisher (Gustav).

Loof, P. A. A. 2001. Nematoda, Secernentea (Tylenchida, Aphelenchida). 246 S. Bd.4/1-1 *In*: Schwoerbel, J. and P. Peter Zwick, (eds.) Begr. v. August Brauer. Süsswasserfauna von Mitteleuropa. 21 Bde. Urban & Fisher; Fisher (Gustav).

Luc, M., R. A. Sikora, and J. Bridge. 1990. Plant-Parasitic Nematodes in Subtropical and Tropical Agriculture. CAB International: Wallingford, United Kingdom.

Mai, W. F., P. G. Mullin, H. H. Lyon, and K. Lowffler. 1996. Pictorial Key to Genera of Plant-Parasitic Nematodes. 5th ed. Cornell University Press: Ithaca, New York.

Malakhov, V. V. V., and D. Hope, (eds.). 1994. Nematodes: Structure, Development, Classification, and Phylogeny. Smithsonian Institution Press: Washington, DC.

Maqbool, M. A., and B. Kerry, (eds.). 1998. Plant Nematode Problems and Their Control in the Near East Region. Bernan Associates: Karachi, Pakistan.

Maqbool, M. A., A. M. Golden, A. Ghaffar, and L. R. Krusberg. 1988. Advances in plant nematology. Shamin Printing Press: Karachi, Pakistan.

Marks, R.J., and B. B. Brodie, (eds.). 1998. Potato Cyst Nematodes. CAB International: Wallingford, United Kingdom.

Matthews, B. E. 1998. An Introduction to Parasitology. Caimbridge University Press: Caimbridge, United Kingdom.

Mayr, E., and P. D. Ashlock. 1991. Principles of Systematic Zoology, 2nd ed. McGraw-Hill, Inc.: New York.

Navon, A., and K. R. S. Ascheer. 2000. Bioassays of Entomopathogenic Microbes and Nematodes. CAB International: Wallingford, United Kingdom.

Nicholas, W. L. 1984. The Biology of Free-living Nematodes. 2nd.ed., Clarendon Press: Oxford, United Kingdom.

Nickle, W. R. (ed.) 1984. Plant and Insect Nematodes, Marcel Dekker: New York.

Nickle, W. R. (ed.) 1991. Manual of Agricultural Nematology. Marcel Dekker: New York.

Norton, D. C. 1978. Ecology of Plant-Parasitic Nematodes. Wiley & Sons, New York.

Pederson, G. A., K. R. Barker, and G. L. Windham, (eds.). 1998. Plant and Nematode Interactions. American Society of Agronomy, Monograph No. 36, Madison Wisconsin.

Perry R. N. and D. J. Wright. 1998. Physiology and Biochemistry of Free-living and Plant-Parasitic Nematodes. CAB International, Wallingford, Oxon, United Kingdom.

Platt, H. M, R. M. Warwick, and P. J. Somerfield, (eds.). 1983-1998. Free-living Marine Nematodes. Published for The Linnean Society of London and the Estuarine and Brackish-Water Sciences Association by Cambridge University Press: Cambridge, United Kingdom.

Poinar, G.O. 1983. The Natural History of Nematodes. Prentice-Hall; Englewood Cliffs: New Jersey.

Poinar, G. O., and H. B. Jansson. 1988. Disease of Nematodes, Volume I. CRC Press: Bocan Raton, Florida.

Poinar, G. O., and H. B. Jansson. 1988. Disease of Nematodes, Volume II. CRC Press: Bocan Raton, Florida.

Riggs, R. D. (ed.) 1982. Nematology in the Southern United States. Southern Cooperative Series Bulletin 276: Fayetteville. Arkansas.

Riggs, R. D., and J. A. Wrather, (eds.). 1992. Biology and Management of Soybean Cyst Nematodes. American Phytopathological Society: St. Paul, Minnesota.

Santos, M. S. N. de A., I. M. de O. Abrantes, D. J. F. Brown, and R. M. Lemos, (eds.). 1997. An Introduction to Virus Vector Nematodes and their Associated Viruses. Universidade de Coimbra: Portugal.

Sasser, J. N. 1989. Plant-Parasitic Nematodes: The Farmer's Hidden Enemy. North Carolina State University, Raleigh, North Carolina.

Sasser, J. N., C. C. Carter, and K. R. Barker. 1985. An Advanced Treatise on *Meloidogyne*. Vol. I Biology and Control. Vol. II Methodology. International *Meloidogyne* Project. North Carolina State University, Raleigh, North Carolina

Sauer, M. R. 1985. A Scanning Electron Microscope Study of Plant and Soil Nematodes. Commonwealth Science Industry Research Organization, Australia.

Sharma, S. B. (ed,) 1998. The Cyst Nematodes. Kluwer Academic Publishers: Dordrecht.

Sharma, R., and G. Swarup. 1988. Pathology of Cyst Nematodes. Malhotra Publishing House: New Delhi.

Shurtleff, M.C., and Averre. C. W. 2000. Diagnosing Plant Diseases Caused by Nematodes. American Phytopathological Society: St. Paul Minnesota.

Siddiqi, M. R. 2000. Tylenchida: Parasites of Plants and Insects. CAB International: Wallingford, United Kingdom.

Simoes, N., N. Boemare, and R. Ehlers, (eds.). 1988. Entompathogenic Nematodes-Pathogenicity of Entomopathogenic Nematodes Versus Insect Defence Mechanisms: Impact on Selection of Virulent Strains. Office for Official Publications of the European Communities.

S'Jacob, J. J., and J. van Bezooijin. 1984. Manual for Practical Work in Nematology. Landbouwhogeschool, Wageningen.

Southey, J. F., (ed.). 1978. Plant Nematology. GD 1, MAFF, HMSO, London.

Southey, J. F. 1986. Laboratory Methods for Work with Plant and Soil Nematodes, 6th ed. Reference Book 402. Ministry of Agriculture, Fisheries, and Food, HMSO: London, United Kingdom.

Starr, J. L., (ed.). 1990. Methods for Evaluating Plant Species for Resistance to Plant Parasitic Nematodes. Society of Nematologists, Hyattsville, Maryland.

Starr, J. R., R. Cook, and J. Bridge. 2002. Plant Resistance to Parasitic Nematodes. CABI Publishing: Wallingford, United Kingdom.

Stirling, G. R., 1991. Biological Control of Plant Parasitic Nematodes. CAB International: Wallingford, United Kingdom.

Stone, A. R., H. M. Platt, and L. F. Khalil. 1983. Concepts in Nematode Systematics.

Taylor, A. L. 1987. Identification and Estimation of Root-knot Nematode Species in Mixed Populations. Florida Department of Agriculture, Division of Plant Industry, Gainesville. Bulletin 12.

Taylor, C. E. and F. J. F. Brown. 1997. Nematode Vectors of Plant Viruses. CAB International: Wallingford, United Kingdom.

Thomson, W. T. 1992. A Worldwide Guide to Beneficial Animals (Insects, Mites, Nematodes) Used for Pest Control Purposes. Thomson Publications: Fresno, California.

Upadhahy, R. K. K. G. Mukerji, and B. P. Chamola. 2000. Crop Diseases, Weeds, and Nematodes, Vol. 1. Kluwer Academic Publishers:

Varma. M. K. 1995. Dictionary of Plant Nematology, Vedams Books International: New Delhi, India.

Veech, J. A. and D. W. Dickson, (eds.). 1987. Vistas on Nematology. Society of Nematologists: Hyattswille, Maryland.

Viglierchio, D. R. 1991. The World of Nematodes. AgAccess: Davis, California.

Websster, J. M. 1972. Economic Nematology. Academic Press: London, United Kingdom.

Weischer, B, and D. J. F. Brown. 2000. An introduction to nematodes: General nematology, A Student's Textbook. Pensoft Publishers: Sofia, Bulgaria.

Wharton, D. A. 1986. A Functional Biology of Nematodes. Johns Hopkins University Press: Baltimore, Maryland.

Whitehead, A. G. 1997. Plant Nematode Control. CAB International: Wallingford, United Kingdom.

Zuckerman, B. M., and D. B. Dusenberry, (eds.) 1980. Nematodes as Biological Models. Academic Press: New York.

Zuckerman, B. M., and R. A. Rhode. 1981. Plant Parasitic Nematodes. Vol. 3. Academic Press: New York.

Zuckerman, B. M., W. F. Mai, and M. B. Harrison. 1990. Laboratory Manual for Plant Nematology. 2nd rev. ed., Massachusetts Agricultural Experiment Station: Amherst. Massachusetts.

CD ROM RESOURCES

Eisenback, J. D. 1998. Taxonomic Database of the Root-knot Nematodes, *Meloidogyne* spp. CAB International: Wallinford, United Kingdom.

Eisenback, J. D. 1999. Glossary of Plant Nematology and Related Terms. CAB International: Wallingford, United Kingdom.

PICTURE COLLECTIONS

Eisenback, J. D., and Ulrich Zunke. 1997. Nemapix Volume 1. Mactode Publications: Blacksburg, Virginia.
Eisenback, J. D., and Ulrich Zunke. 1999. Nemapix Volume 2. Mactode Publications: Blacksburg, Virginia.
Eisenback, J. D., and Ulrich Zunke. 2002. Nemapix Volume 3. Mactode Publications: Blacksburg, Virginia.

SLIDE SERIES

Society of Nematologists, Slide Set Nematology, 100 Slides. T.A. Niblack, Department of Plant Pathology, University of Missouri, Columbia, Missouri.
Högger, C.H. FAP Dia-Serien 1986: Nr. 3 Nematoden: Formenvielfalt und Antagonisten; Nr. 4 Nematoden an Getreide; Nr. 5 Nematoden an Kartoffeln; Nr. 6 Nematoden an Zucker- und Futterrüben, je 20 Dias, Legende, FAL Zürich-Reckenholz.
De Grisse, A. 1976. Rasterelektronenmikroskop-Aufnahmen (SEM) von Nematoden. 113 Dias. C. De Grisse-Helleybuck, Schoonmeerstr. 37, B-9000 Ghent, Belgien.
Society of Nematologists. 1990. Slide Set: Parasites and Predators of Plant-Parasitic Nematodes. 60 slides. Society of Nematologists Biological Control Committee.

VIDEOS

Eisenback, J. D., and E. C. McGawley. 1996. A Videocassette for Teaching the Identification of the Most Common Genera of Plant-Parasitic Nematodes. Mactode Publications: Blacksburg, Virginia.

Appendix 6

References to Nematological Keys

Aglenchus Andrássy, 1954

Andrássy, I. 1980. The genera and species of the family Tylenchidae Oerley, 1880 (Nematoda). Acta Zoologica Academiae Scientiarum Hungaricae 26:1-20.

Amplimerlinius Siddiqi, 1976

Brzeski, M. W. 1985. Contribution to the knowledge of Polish plant parasitic nematodes (Nematoda). 7. Ten species new for Polish fauna. Fragmenta Faunistica 29:275-288.

Shaw, S. P., and E. Khan. 1992. Tylenchorhynchidae Golden, 1971 (Nematoda: Tylenchida) in India: 1. *Amplimerlinius sikkimensis* sp. n. from Sikkim with key to *Amplimerlinius* species and *Tylenchorhynchus shivanandi* sp. n. from Nagaland. Bulletin of Entomology, New Delhi 33:7-13.

Anguina Scopoli, 1777 (after Chizhov and Subbotin, 1990)

Chizhov, V. N., and S. A. Subbotin. 1990. Phytoparasitic nematodes of subfamily Anguininae (Nematoda,Tylenchida). Morphology, trophic specialization, taxa. Zoologicheskii Zhurnal 69:15-26. (In Russian).

Aorolaimus Sher, 1963

Baujard, P., P. Castillo, M. E. Doucet, B. Martiny, and D. Mounport. 1994. Taxonomic studies on the genus *Aorolaimus* Sher, 1963 (Nemata: Hoplolaimidae) I. Bibliographic analysis and tentative key to species. Fundamental and Applied Nematology 17:103-115.

Aphelenchoides Fischer, 1894

Sanwal, K. C. 1961. A key to the species of the nematode genus *Aphelenchoides* Fischer, 1894. Canadian Journal of Zoology 39:143-148.

Hunt, D. J. 1993. Aphelenchida, Longidoridae, and Trichodoridae: Their systematics and bionomics. Wallingford, UK: CAB International.

Shahina, F. 1996. A diagnostic compendium of the genus *Aphelenchoides* Fischer, 1894 (Nematoda: Aphelenchida) with some new records of the group from Pakistan. Pakistan Journal of Nematology 14:1-32.

Aphelenchus Bastian, 1865

Nama, H. S., and Soni, G. R. 1981. Taxonomy of some species of the genus *Aphelenchus* Bastian, 1865 (Aphelenchoidea) with a key. Proceeding of the Indian Academy of Parasitology 2:107-110.

Anderson, R. V., and D. J. Hooper. 1980. Diagnostic value of vagina structure in the taxonomy of *Aphelenchus* Bastian, 1865 (Nematoda: Aphelenchidae) with a description of *A. (Anaphelenchus) isomerus* n. subgen., n. sp. Canadian Journal of Zoology 58:924-928.

Bakernema Wu, 1964

Ebsary, B. A. 1982. *Bakernema yukonense* n. sp. (Nematoda: Criconematidae) with keys to the species of *Criconemella* and *Discocriconemella*. Canadian Journal of Zoology 60:3033-3047.

Basiria Siddiqi, 1959

Karegar, A., and E. Geraert. 1997. The genus *Basiria* Siddiqi, 1959 (Nematoda: Tylenchida). II. Species with four lateral lines and anterior median bulb. Nematologica 43:383-406.

Karegar, A., and E. Geraert. 1998. The genus *Basiria* Siddiqi, 1959 (Nemata: Tylenchida) IV. General discussion, genus diagnosis and key to the species. Nematologica 44:1-13.

Belonolaimus Steiner, 1949

Rau, G. J. 1963. Three new species of *Belonolaimus* (Nematoda: Tylenchida) with additional data on *B. Longicaudatus* and *B. gracilis*. Proceedings of the Helminthological Society of Washington 30:119-128.

Bursaphelenhus Fuchs, 1937

Tarjan, A. C., and C. B. Aragon. 1982. An analysis of the genus *Bursaphelenchus* Fuchs, 1937. Nematropica 12:121-144.

Yin, K., Y. Fang, and A. C. Tarjan. 1988. A key to species in the genus *Bursaphelenchus* with a description of *Bursaphelenchus hunanensis* sp. n. (Nematoda: Aphelenchoididae) found in pine wood in Hunan Province, China. Proceedings of the Helminthological Society of Washington 55:1-11.

Boleodorus Throne, 1941

Khan, E. 1963. *Boleodorus mirus* n. sp. (Tylenchida: Boleodorinae n. subfam.) from Kufri, Simla (H. P.) India, with a key to the species of the genus *Boleodorus* Thorne, 1941. Zool. Anz. 173:336-341.

Geraert, E. 1971. Observations on the genera *Boleodorus* and *Boleodoroides* (Nematoda: Tylenchidae). Nematologica 17:263-276.

Brzeski, M. W., and M. R. Sauer. 1982. Scanning electron micrography of some Tylenchidae and Boleodoridae and reappraisal of the Boleodoridae. Nematologica 28:437-446.

Cephalenchus (J. B. Goodey, 1962) Golden, 1971

Sultan, M. S., and M. S. Jairajpuri. 1981. Two new species of the genus *Cephalenchus* (Goodey, 1962) Golden, 1971 with a key to species. Indian Journal of Nematology 11:165-171.

Mizukubo, T., and N. Minagawa. 1985. Taxonomic study of the genus *Cephalenchus* (Nematoda:Tylenchida) from Japan. Descriptions of three new species and records of *C. planus* Siddiqui & Khan with a key to species. Japanese Journal of Nematology 15:26-40.

Raski, D. J., and E. Geraert. 1986. Descriptions of two new species and other observations on the genus *Cephalenchus* Goodey, 1962 (Nemata: Tylenchida). Nematologica 32:56-78.

Coslenchus Siddiqi, 1978

Brzeski, M. W. 1978. Taxonomic notes on *Coslenchus* Siddiqi, 1978 (Nematoda: Tylenchidae). Annales Zoologici 28:417-436.

Siddiqi, M. R. 1980. Six new species of *Coslenchus* Siddiqi, 1978 (Nematoda: Tylenchidae). Nematologica 26:432-447.

Andrássy, I. 1982. The genera and species of the family Tylenchidae Orley, 1880 (Nematoda). The genus *Coslenchus* Siddiqi, 1978. Acta Zoologica Academiae Scientiarum Hungaricae 28:193-232.

Geraert, E., and D. J. Raski. 1988. Study of some *Aglenchus* and *Coslenchus* species (Nematoda:Tylenchida). Nematologica 34:6-46.

Criconema Hofmanner & Menzel, 1914

Golden, A., and W. Friedman. 1964. Some taxonomic studies on the genus *Criconema* (Nematoda: Criconematidae). Proceedings of the Helminthological Society of Washington 31:47-59.

Yeates, G. W., P. A. A. A. Loof, and W. M. Wouts. 1997. Criconematidae (Nematoda: Tylenchida) from the New Zealand region: analysis of and key to *Criconema* (*Nothocriconemella*) species. New Zealand Journal of Zoology 24: 153-162.

Criconemella (Mesocriconema) de Grisse & Loof, 1965 and related genera

Raski, D. J. 1952. On the morphology of *Criconemoides* Taylor, 1936, with descriptions of six new species (Nematoda: Criconematidae). Proceedings of the Helminthological Society of Washington 19:85-99.

De Grisse, A., and P. A. A. A. Loof. 1965. Revision of the genus *Criconemoides* (Nematoda). Mededelingen Landbouwhogeschool en Opzoekingsstations van de Staat te Gent 30:577-603. (Landmark work, keys to species of the various Criconemoides segregates)

Tarjan, A. C. 1966. A compendium of the genus *Criconemoides* (Criconematidae: Nemata). Proceedings of the Helminthological Society of Washington 33:109-125.

Raski, D. J., and A. M. Golden. 1966. Studies on the genus *Criconemoides* Taylor, 1936 with descriptions of eleven new species and *Bakernema variabile* n. sp. (Criconematidae: Nematoda). Nematologica 11:501-565.

Mehta, U. K., and D. J. Raski. 1971. Revision of the genus *Criconema* Hofmanner and Menzel, 1914 and other related genera (Criconematidae: Nematoda). Indian Journal of Nematology 1:145-198. (includes *Ogma*, *Crossonema* and others).

Andrássy, I. 1979. Revision of the subfamily Criconematinae Taylor, 1936 (Nematoda). Opuscula Zoologica (Budapest) 16:11-57. (Many keys to genera and species)

Ebsary, B. A. 1979. Description of *Criconemoides pleriannulatus* n. sp. (Nematoda: Criconematidae) and a key to species of *Criconemoides*. Canadian Journal of Zoology 57:1-5.

Luc, M., and D. J. Raski. 1981. Status of the genera *Macroposthonia*, *Criconemoides*, *Criconemella*, and *Xenocriconemella* (Criconematidae: Nematoda). Revue de Nématologie 4:3-21.

Hashim, Z. 1984. Re-diagnosis and a key to the species of *Neolobocriconema* Mehta & Raski, 1971 (Nematoda: Tylenchida), with a description of *N. orlarum* n. sp. from Jordan. Systematic Parasitology 6:69-73.

Loof, P. A. A., and De Grisse. 1989. Taxonomic and nomenclatorial observation of the genus *Criconemella* De Grisse & Loof, 1965 *sensu* Luc & Raski (Criconematidae). Mededlingen Landbouwhogeschool Wageningen, Rijksuniversiteit Gent 54:53-74.

Cryphodera Colbran, 1966

Karssen, G., and A. Van Aelst. 1999. Description of *Cryphodera brinkmani* n. sp. (Nematoda: Heteroderidae), a parasite of *Pinus thunbergii* Parlatore from Japan, including a key to the species of *Cryphodera* Colbran, 1966. Nematology 1:121-130.

Ditylenchus Filip'ev, 1936

Brzeski, M. W. 1991. Review of the genus *Ditylenchus* Filipjev, 1936 (Nematoda: Anguinidae). Revue de Nématologie 14:9-59.

Dolichodorus Cobb, 1914

Loof, P. A. A., and R. D. Sharma. 1975. *Dolichodorus minor* n. sp. (Nematoda: Dolichodoridae) with a key to the genus *Dolichodorus*. Revista Theobroma (Brazil) 5:35-41.

Lewis, S. A., and A. M. Golden. 1981. Description and SEM observation of *Dolichodorus marylandicus* n. sp. with a key to species of *Dolichodorus*. Journal of Nematology 13:128-134.

Smart, G. C., Jr., N. B. Khuong. 1985. *Dolichodorus miradvulvus* n. sp. (Nematoda: Tylenchida) with a key to species. Journal of Nematology 17:29-37.

Eutylenchus Cobb, 1913

Brzeski, M. W. 1996. On the genus *Eutylenchus* Cobb, 1913 (Nematoda: Tylenchidae). Nematologica 42:1-8.

Filenchus Andrássy, 1954

Raski, D. J., and E. Geraert. 1986. Review of the genus *Filenchus* Andrássy, 1954 and descriptions of six new species (Nemata: Tylenchidae). Nematologica 32:265-311. (Synonymy of *Ottolenchus*, *Dactylotylenchus*, *Lambertia*, *Duosulcius*, *Zanenchus*, and *Discotylenchus* with *Filenchus*)

Geocenamus Thorne & Malek, 1968

Brzeski, M. W. 1991. Taxonomy of *Geocenamus* Thorne & Malek, 1968 (Nematoda: Belonolaimidae). Nematologica 37:125-173.

Brzeski, M. W. 1992. Species omitted in *Geocenamus* review. Nematologica 38:255 (Short communication).

Globodera Skarbilovich, 1959

Wouts, W. M. 1984. *Globodera zelandica* n. sp. (Nematoda: Heteroderidae) from New Zealand, with a key to the species of *Globodera*. New Zealand Journal of Zoology 11:129-135.

Gracilacus Raski, 1962

Raski, D. J. 1975. Revision of the genus *Paratylenchus* Micoletzky, 1922 and descriptions of new species. Part III of three parts-*Gracilacus*. Journal of Nematology 8: 97-115.

Helicotylenchus Steiner, 1945

Sher, S. A. 1966. Revision of the Hoplolaiminae (Nematoda). VI. *Helicotylenchus* (Steiner, 1945). Nematologica 12: 1-56.

Anderson, R. V. 1979. A supplemental key to species of *Helicotylenchus* Steiner, 1945 (Nematoda: Hoplolaimidae) described since 1972 and a description of *H. oscephalus* n. sp. Canadian Journal of Zoology 57:337-342.

Anderson, R. V., and E. S. Eveleigh. 1982. Description of *Helicotylenchus amplius* n. sp. and a key to the Canadian species of the genus (Nematoda: Hoplolaimidae). Canadian Journal of Zoology 60:318-321.

Boag, B., and M. S. Jairajpuri. 1985. *Helicotylenchus scoticus* n. sp. and a conspectus of the genus *Helicotylenchus* Steiner, 1945 (Tylenchida: Nematoda). Systematic Parasitology 7:47-58. (Compendium of 154 species)

Fotedar, D. N. and V. Kaul. 1985. A revised key to the species of the genus *Helicotylenchus* Steiner, 1945 (Nematoda: Rotylenchoidinae). Indian Journal of Nematology 15:138-147. (Approximately 125 species in this key)

Firoza, K., and M. A. Maqbool, M. A. 1994. A diagnostic compendium of the genus *Helicotylenchus* Steiner, 1945 (Nematoda: Hoplolaimidae). Pakistan Journal of Nematology 12:11-50.

Hemicriconemoides Chitwood & Birchfield, 1957

Dasgupta, D. R., D. J. Raski, and S. D. Van Gundy. 1969. Revision of the genus *Hemicriconemoides* Chitwood & Birchfield, 1957 (Nematoda: Criconematidae). Journal of Nematology 1: 126-145.

Germani, G., and M. Luc. 1970. Contribution a l'etude du genre *Hemicriconemoides* Chitwood & Birchfield, 1957 (Nematoda: Criconematidae). Cahiers ORSTOM Serie Physiologie des Plantes Tropicales Cultivees 11:133-150.

Esser, P. R. and Volvas, N. 1990. A diagnostic compendium to the genus *Hemicriconemoides* (Tylenchida: Criconematidae). Proceedings of the Soil and Crop Science Society of Florida 49:211-219.

Hemicycliophora de Man, 1921

Schoemaker, R. L. P. 1967. *Hemicycliophora nyanzae* n. sp. found in East Africa with a key to the species of *Hemicycliophora* De Man, 1921. Nematologica 3: 541-546.

Loof, P. A. A. 1968. Taxonomy of *Hemicycliophora* species from west and central Europe (Nematoda: Criconematoidea). Mededlingen Landbouwhogeschool Wageningen 68-14:1-43.

Brzeski, M. W. 1974. Taxonomy of Hemicycliophorinae (Nematoda, Tylenchida). Zeszyty Problemowe Postepow Nauk Rolniczych 154:237-330.

Loof, P. A. A. A. 1976. The genera *Hemicycliophora* De Man, 1921 and Caloosia Siddiqi & Goodey, 1963 (Criconematoidea). Mededelingen van de Faculteit Landbouwwetenschappen Rijksuniversiteit Gent 41:1023-1029.

Eroshenko, A. 1976. Paraziticheski kornevye nematody. Semeistvo Hemicycliophoridae, Leningrad: Nauka Leningradskoe otd-nie. 79p.

Brzeski, M. W., and T. S. Ivanova. 1978. Taxonomic notes on *Hemicycliophora* de Man (Nematoda: Hemicycliophoridae). Nematologia Mediterranea 6:147-162.

Siddiqi, M. R. 1980. Taxonomy of the plant nematode superfamily Hemicycliophoroidea, with a proposal for Criconematina, a new suborder. Revue de Nématologie 3:179-199.

Van Den Berg, E. 1987. *Hemicycliophora* species from the western Cape Province with a key to the South African species (Hemicycliophoridae: Nematoda). Phytophylactica 19:303-307.

Heterodera Schmidt, 1871

Mulvey, R. H. 1972. Identification of *Heterodera* cysts by terminal and cone top structures. Canadian Journal of Zoology 50:1277-1292.

Golden, A. M. 1986. Morphology and identification of cyst nematodes. Pp. 23-45 *in* F. Lamberti and C. E. Taylor, eds. Cyst nematodes. NATO ASI Series. Series A: Life Sciences Vol. 121. New York: Plenum Press.

Wouts, W. M., A. Schoemaker, D. Sturhan, and P. R. Burrows. 1995. *Heterodera spinicauda* sp. n. (Nematoda: Heteroderidae) from mud flats in the Netherlands, with a key to the species of the *H. avenae* group. Nematologica 41:575-583.

Hirschmaniella Luc and Goodey, 1964

Ebsary, B. A., and R. V. Anderson. 1982. Two new species of *Hirschmaniella* Luc and Goody, 1963 (Nematoda: Pratylenchidae) with a key to the nominal species. Canadian Journal of Zoology 60:530-534.

Sivakumar, C. V., and E. Khan. 1982. Description of *Hirschmaniella kaverni* sp. n. (Radopholidae: Nematoda) with a key for identification of *Hirschmaniella* spp. Indian Journal of Nematology 12:86-90.

Hoplolaimus Daday, 1905

Sher, S. A. 1963. Revision of the Hoplolaiminae (Nematoda). II. *Hoplolaimus* Daday, 1905 and *Aorolaimus* n. gen. Nematologica 9:267-295.

Jairajpuri, M. S., and Q. H. Baqri. 1973. Nematodes of high altitudes in India. I. four new species of Tylenchida. Nematologica 19:19-30.

Anderson, R. V. 1983. Morphological characteristics of *Hoplolaimus indicus* Sher,1963 in Canada, a parasite of wild rice. Journal of Nematology 15:366-369.

Handoo, Z. A., and A. M. Golden. 1992. A key and diagnostic compendium tothe species of the genus *Hoplolaimus* Daday, 1905 (Nematoda: Hoplolaimidae). Journal of Nematology 15:45-53.

Hoplotylus s'Jacob, 1960

Bernard, E. C., and T. L. Niblack. 1982. Review of *Hoplotylus* s'Jacob (Nematoda: Pratylenchidae). Nematologica 28:101-109.

Longidorus Micoletzky, 1922 (Filip'ev, 1934; Thorne, 1934) and
Paralongidorus Siddiqi, Hooper & Khan, 1963

Romanenko, N. D. 1978. A polytomic key for the identification of species of *Longidorus* Micoletzky, 1922. Fitogel'mintologicheskie issledovaniya. Moscow, "Nauka." 111-114.

Chen, Q. W., D. J. Hooper, P. A. A. Loof, and J. H. Xu. 1997. A revised polytomous key for the identification of species of the genus *Longidorus* Micoletzky, 1922 (Nematoda: Dorylaimoidea). Fundamental and Applied Nematology 20:15-28.

Escuer, M., and M. Arias. 1997. *Paralongidorus iberis* sp. n. from Spain with a polytomous key to the species of the genus *Paralongidorus* Siddiqi, Hooper & Khan, 1963 (Nematoda: Longidoridae). Fundamental and Applied Nematology 20:135-148.

Taylor, C. E., and D. J. F. Brown. 1997. Nematode vectors of plant viruses. Wallingford, UK: CAB International. 286 p.

Loof, P. A. A., and Q. W. Chen. 1999. A revised polytomous key for the identification of species of the genus *Longidorus* Micoletzky, 1922 (Nematoda: Dorylaimoidea). Supplement I. Nematology 1:55-59.

Malenchus Andrássy, 1968

Andrássy, I. 1981. The genera and species of the family Tylenchidae Oerley, 1880 (Nematoda). The genus *Malenchus* Andrássy, 1968.). Acta Zoologica Academiae Scientiarum Hungaricae 27:1-47.

Geraert, E., and D. J. Raski. 1986. Unusual *Malenchus* species (Nemata: Tylenchidae). Nematologica 32:27-55.

Meloidogyne Goeldi, 1889

Taylor, A. L., and J. N. Sasser. 1978. Biology, identification and control of root-knot nematodes (*Meloidogyne* species). Raleigh: North Carolina State University Graphics.

Eisenback, J. D., H. Hirschmann, J. N. Sasser, and A. C. Triantaphyllou. 1981. A guide to the four most common species of root-knot nematodes (*Meloidogyne* spp.), with a pictorial key. Raleigh: North Carolina State University.

Ebsary, B. A., and E. S. Eveleigh. 1983. *Meloidogyne aquatilis* n. sp. (Nematoda: Meloidogynidae) from Spartina pectinata with a key to the Canadian species of Meloidogyne. Journal of Nematology 15:349-353.

Hewlett, T. E., and A. C. Tarjan. 1983. Synopsis of the genus *Meloidogyne* Goeldi, 1887. Nematropica 13:79-102.

Jepson, S. B. 1987. Identification of root-knot nematodes (*Meloidogyne* species). Wallingford, UK: CAB International. 265 p.

Eisenback, J. D., and H. H. Triantaphyllou. 1991. Root-knot nematodes: *Meloidogyne* species and races. Pp. 191-274 *in* W. R. Nickel, ed. Manual of agricultural nematology. New York: Marcel Dekker.

Merlinius Siddiqi, 1970

Sultan, M. S., I. Singh, and P. K. Sakhuja. 1988. Plant parasitic nematodes of Punjab, India V. *Merlinius communica* n. sp. with key to species of the genus and *Telotylenchus elongatus* n. sp. Indian Journal of Nematology 18:44-48.

Sultan, M. S., I. Singh, and P. K. Sakhuja. 1988. Addendum to the paper "Plant parasitic nematodes of Punjab, India. V. *Merlinius communica* n. sp. with a key to the species of the genus and *Telotylenchus elongatus* n. sp. Indian Journal of Nematology 18:44-48.

Nacobbus Thorne and Allen, 1944

Sher, S. A. 1970. Revision of the genus *Nacobbus* Thorne and Allen, 1944 (Nematoda: Tylenchoidea). Journal of Nematology 2:228-235.

Nagelus Malek & Thorne, 1968

Powers, T. O., J. G. Baldwin, and A. H. Bell. 1983. Taxonomic limits of the genus *Nagelus* (Throne and Malek, 1968) Siddiqi, 1979 with a description of *Nagelus borealis* n. sp. from Alaska. Journal of Nematology 15:582-593. (Tabular information)

Neopsilenchus Throne and Malek, 1968

Khan, F. A., and A. M. Khan. 1975. Three new species of *Neopsilenchus* Thorne & Malek, 1968 (Nematoda: Psilenchinae). Indian Journal of Nematology 5:15-21.

Sultan, M. S., I. Singh, P K. Sakhuja. 1987. Plant parasitic nematodes of the Punjab, II. *Neopsilenchus longicaudatus* n. sp. (Nematoda: Tylenchidae). Indian Journal of Nematology 17:330-332.

Shahina, F., and Maqbool, M. A. 1990. Studies on the genus *Neopsilenchus* (Nematoda: Tylenchidae) with description of *Acusilenchus* n. subgen and three new species. Pakistan Journal of Nematology. 8:49-63.

Ogma Southern, 1914

Minagwa, N. 1993. Taxonomic studies of Criconematidae (Nematoda: Tylenchida) of Japan IV *Ogma*: part 2. Bulletin National Agro. Environmental Science 9:53-152.

Van Den Berg, E., and P. Queneherve. 1995. *Ogma toparti* sp.n. and two known Criconematoidea from the French Caribbean (Nemata: Tylenchina). Fundamental and Applied Nematology 18:361-369.

Ottolenchus Husain & Khan, 1967 (Wu, 1970) Golden, 1971 synonymized this subgenus with ***Aglenchus***, Andrássy, 1954

Brzeski, M. W. 1982. Taxonomy of Ottolenchus Husain & Khan, and a description of *Coslenchus polonicus* sp. n. (Nematoda: Tylenchidae). Revue de Nématologie 5:71-77.

Maqbool, M. A., and F. Shahina. 1985. Two new species of the genus *Ottolenchus* Husain & Khan, 1967 (Nematoda: Tylenchidae) with a key to the species and observation on *O. facultativus* (Szczygiel, 1969) Brzeski, 1982 from Pakistan. Revue de Nématologie 8:329-333.

Sumenkova, N. I. 1987. Taxonomy of phytonematodes of the genus *Ottolenchus* (Nematoda: Tylenchidae) and a description of two new species of the genus. Zoologicheskii Zhurnal 66:1805-1811.

Pararotylenchus Baldwin and Bell, 1981

Baldwin and Bell, A. H. 1981. *Pararotylenchus* n. gen. (Pararotylenchinae n. subfam.: Hoplolaimidae) with six new species and two new combinations. Journal of Nematology 13:111-128.

Paratrophurus Arias, 1970 (*Histotylenchus* Siddiqi, 1971 and Telotylenchoides Siddiqi, 1971 are synonymous with *Paratrophurus*, according to Ebsary, 1991.)

Paratylenchus Micoletzky, 1922 and ***Gracilacus*** Raski, 1962

Geraert, E. 1965. The genus *Paratylenchus*. Nematologica 11:301-334. (Old but contains lots of good information and comparative drawings.)

Solov'eva, G. I. 1972. Parasitic nematodes of woody and herbaceous plants. Akadeiya Nauk SSSR (Lenningrad, Russia: Nauka Publishers).

Raski, D. J. 1975. Revision of the genus *Paratylenchus* Micoletzky, 1922 and descriptions of new species. Part I of 3 parts. Journal of Nematology 7:15-34.

Raski, D. J. 1975. Revision of the genus *Paratylenchus* Micoletzky, 1922 and descriptions of new species. Part II of three parts. Journal of Nematology 7:274-295.

Esser, R. P. 1992. A diagnostic compendium to species included in Paratylenchinae Thorne, 1949 and Tylenchocriconematinae Raski & Siddiqui, 1975 (Nematoda: Criconematoidea). Nematologica 38: 146-163.

Peltamigratus Sher, 1964

Mulk, M. M., and M. R. Siddiqi. 1982. Three new species of Hoplolaimid nematodes from South America. Indian Journal of Nematology 12:124-131.

Bitencourt , C., and C. S. Huang. 1986. Brazilian *Peltamigratus* Sher, 1964 (Nematoda: Hoplolaimidae), with description of six new species. Revue de Nématologie 9:3-24.

Rashid, F, E. Geraert , and R. D. Sharma. 1987. The genus *Peltamigratus* Sher, 1964 with description of two new species (Nematoda: Hoplolaimidae). Revue de Nématologie 10:3-21.

Fortuner, R. 1987. A reappraisal of Tylenchina (Nemata). 8. The family Hoplolaimidae Filip'ev, 1934. Revue de Nématologie 10:219-232. (*Peltamigratus* was placed in synonymy with *Aorolaimus*.)

Pratylenchoides Winslow, 1958

Ryss, A. Y. 1980. *Pratylenchoides ivanovae* sp. n. (Nematoda: Pratylenchidae) and key to species of the genus *Pratylenchoides*. Parazitologiya 14:516-520.

Baldwin, J. G., et al. 1983. Contribution to the study of the genus *Pratylenchoides* Winslow (Nematoda: Tylenchidae). Revue de Nématologie 6:111-125.

Pratylenchus Filip'ev, 1936

Thorne, G. 1961. Principles of Nematology. New York: McGraw-Hill. (Still a good reference for becoming acquainted with the primary identifying features)

Loof, P. A. A. A., 1978. The genus *Pratylenchus* Filipjev, 1936 (Nematoda: Pratylenchidae): a review of its anatomy, morphology, distribution, systematics and identification. Vaxtskyddsrapporter 5:1-50. (Outstanding key)

Ryss, A. Yu. 1988. Kornevye parazitícheskie nematody semeistva Pratylenchidae (Tylenchida) mirovoi fauny. Lenningrad, Nauka. (Russian text) (See software.

Frederick, J. J., and A. C. Tarjan. 1989. A compendium of the genus *Pratylenchus* Filipjev, 1936 (Nemata: Pratylenchidae). Revue de Nématologie 12:243-256.

Handoo, Z. A., and A. M. Golden. 1989. A key and diagnostic compendium to the genus *Pratylenchus* Filipjev, 1936 (lesion nematodes). Journal of Nematology 21:202-218.

Café Filho, A. C., and C. S. Huang. 1989. Description of *Pratylenchus pseudofallax* n. sp. with a key to the species of the genus *Pratylenchus* Filipjev, 1936 (Nematoda: Pratylenchidae). Revue de Nématologie 12:7-15.

Psilenchus de Man, 1921.

Kheiri, A. 1970. Two new species in the family Tylenchidae (Nematoda) from Iran, with a key to *Psilenchus* De Man, 1921. Nematologica 16:359-368.

Brzeski, M. W. 1989. Notes on the genus *Psilenchus* de Man, 1921, with description of *P. klingleri* sp. n. (Nematoda: Tylenchidae). Annales Zoologici 43:51-69.

Quinisulcius Siddiqi, 1971

Maqbool, M. A. 1992. Description of *Quinisulcius solani* n. sp. (Nematoda: Tylenchorhynchidae) with a key to the species and data on *Scutylenchus koreanus* from Pakistan. Journal of Nematology 14:221-225.

Radopholus Thorne, 1949

Sher, S. A. 1968. Revision of the genus *Radopholus* Thorne, 1949 (Nematoda: Tylenchidae). Proceedings of the Helminthological Society of Washington 35:219 237.

Rhadinaphelenchus J. B. Goodey, 1960
Monotypic: *R. cocophilus* (Cobb, 1919) J. B. Goodey, 1960

Rotylenchus Filip'ev, 1936

Sher, S. A. 1965. Revision of the Hoplolaiminae (Nematoda) V. *Rotylenchus* Filipjev, 1936. Nematologica 11:173-198.

Boag, B., and D. J. Hooper. 1981. *Rotylenchus ouensensis* n. sp. (Nematoda: Hoplolaimidae) from the British Isles. Systematic Parasitology 3:119-125.

Geraert, E., and S. Barooti. 1996. Four *Rotylenchus* from Iran, with a key to the species. Nematologica 42:503-520.

Castillo, P., N. Vovlas, A. Gomez-Barcina and F. Lamberti. 1993. The plant parasitic nematode *Rotylenchus*, (a monograph). Supplement to Nematologia Mediterranea 21. 200 p.

Rotylenchulus Linford & Oliverra, 1940

Robinson, A. F., R. N. Inserra, E. P. Caswell-Chen, N. Vovlas, and A. Troccoli. 1997. *Rotylenchulus* species: identification, distribution, host ranges, and resistance. Nematropica 27:128-180. (Useful descriptions – not a key. See Nickle, Manual of Agricultural Nematology.)

Scutellonema Andrássy, 1958

Sher, S. A. 1963. Revision of the Hoplolaiminae (Nematoda) III. *Scutellonema* Andrássy, 1958. Nematologica 9:421-443.

Van Den Berg, E., and J. Heyns. 1973. South African Hoplolaiminae 2. The genus *Scutellonema* Andrassy, 1958. Phytophylactica 5:23-40.

Germani, G., J. G. Baldwin, A. H. Bell, X. Y. Wu. 1986. Revision of the genus *Scutellonema* Andrássy, 1958 (Nematoda: Tylenchida). Revue de Nématologie 8:289-320.

Telotylenchus Siddiqi, 1960

Trophurus Loof, 1956

Kleynhans, K. P., and P. Cadet. 1994. *Trophurus deboeri* n. sp. from sugarcane soil in Barbados and key to the species of the genus *Trophurus* Loof, 1956 (Nematoda: Belonolaimidae). Fundamental and Applied Nematology 17:225-230.

Tylenchorhynchus Cobb, 1913 and related forms

Tarjan, A. C. 1973. A synopsis of the genera and species in the Tylenchorhynchinae (Tylenchoidea, Nematoda). Proceedings of the Helminthological Society of Washington 40:123-144.

Jairajpuri, M. S., and D. J. Hunt. 1984. The taxonomy of Tylenchorhynchinae (Nematoda: Tylenchida) with longitudinal lines and ridges. Systematic Parasitology 6:261-268. (Keys to *Dolichorhynchus*, *Neodolichorhynchus*, *Tessellus*, and to genera of subfamily)

Fortuner, R., and M. Luc. 1987. A reappraisal of Tylenchina (Nemata). 6. The family Belonolaimidae Whitehead, 1960. Revue de Nématologie 10:183-202. (Five genera synonymized with *Tylenchorhynchus*)

Brzeski, M. W., and C. M. Dolinski. 1998. Compendium of the genus *Tylenchorhynchus* Cobb, 1913 *sensu lato* (Nematoda: Belonolaimidae). Russian Journal of Nematology 6:189-199. (Synonymy of six genera with *Tylenchorhynchus*)

There is no key for identification of species after rearrangement of *Tylenchorhynchus* and related genera by Fortuner and Maggenti, 1987, and many species have been described after Tarjan, 1973. A tabular key [Brzeski and Dolinski, 1998] may be more practical for identification.

Tylenchus Bastian, 1865 and related forms

Andrássy, I. 1954. Revision der gattung *Tylenchus* Bastian, 1865 (Tylenchidae, Nematoda). Acta Zoologica Academiae Scientiarum Hungaricae 1:5-42. (Foundation manuscript for studying Tylenchidae)

Bello, A. 1973. Étude taxonomique du genre Tylenchus Bastian 1865 (Nematoda: Tylenchinae). Nematologia Mediterranea 1:1-7.

Andrássy, I. 1979. The genera and species of the family Tylenchidae Oerley,1880 (Nematoda). The genus *Tylenchus* Bastian, 1865. Acta Zoologica Academiae Scientiarum Hungaricae 25:1-33.

Siddiqi, M. R. 1979. Seven new species in a new nematode subfamily, Duosulciinae n. subfam., is proposed to accommodate species with only two incisures (or a single ridge) in the lateral field. Nematologica 25:215-236. (Key to the six genera)

Brzeski, M. W., and M. R. Sauer. 1982. Scanning electron microscopy of some Tylenchidae and Boleodoridae and reappraisal of the Boleodoridae. Nematologica 28:437-446

Sumenkova, N. I. 1984. Survey of generic and suprageneric categories of nematodes in the family Tylenchidae Orley, 1880. Taksonomiya I biologiya fitogel'mintov. 1984. 132-144. Moscow: Nauka. (A key to genera, in Russian.)

Tylenchulus Cobb, 1913

Inserra, R. N., N. Vovlas, J. H. O'Bannon, and R. P. Esser. 1988. *Tylenchulus graminis* n. sp. and *T. palustris* n. sp. (Tylenchulidae), from native flora of Florida, with notes on *T. semipenetrans* and *T. furcus*. Journal of Nematology 20: 266-287.

Trichodorus Cobb, 1913 and related genera

Esser, R. P. 1971. A compendium of the genus *Trichodorus* (Dorylaimoidea: Diphtherophoridae). Soil and Crop Science 31:244-253.

Siddiqi, M. R. 1973. Systematics of the genus *Trichodorus* Cobb, 1913 (Nematoda: Dorylaimida), with descriptions of three new species. Nematologica 19:259-278. (A key to the subgenera and species of *Paratrichodorus* is given)

Decraemer, W. 1980. Systematics of the Trichodoridae (Nematoda) with keys to their species. Revue de Nématologie 3:81-99.

Rashid, F., D. de Waele, and A. Coomans. 1985. Trichodoridae from Brazil. Nematologica 31:289-320. (Emended diagnosis and key for *Allotrichodorus*)

Waele, D. de, and M. W. Brzeski. 1995. *Trichodorus altaicus* sp. n. (Nematoda: Trichodoridae) and a key to the species of the genus *Trichodorus*. Fundamental and Applied Nematology 18:181-187.

Decraemer, W. 1995. The family Trichodoridae: Stubby root and virus vector nematodes. Dordrecht, The Netherlands: Kluwer Academic Publishers. 360 p.

Taylor, C. E., and D. J. F. Brown. 1997. Nematode vectors of plant viruses. Wallingford, UK: CAB International. 286 p.

Decraemer, W., and P. Baujard. 1998. A polytomous key for the identification of species of the family Trichodoridae Thorne, 1935 (Nematoda: Triplonchida). Fundamental and Applied Nematology 21:37-62.

Xiphinema Cobb, 1913

Lamberti, F., and T. Beleve-Zacheo. 1979. Studies on *Xiphinema americanum* sensu lato with descriptions of fifteen new species (Nematoda, Longidoridae). Nematologia Mediterranea 7:51-106.

Loof, P. A. A., and M. Luc. 1990. A revised polytomous key for the identification of species of the genus *Xiphinema* Cobb, 1913 (Nematoda: Longidoridae) with exclusion of the *X. americanum*-group). Systematic Parasitology 16:35-66.

Lamberti, F., and M. Carone. 1991. A dichotomous key for the identification of species of *Xiphinema* (Nematoda: Dorylaimida) with the *X. americanum*–group. Nematologia Mediterranea 19:341-348.

Loof, P. A. A., and M. Luc. 1993. A revised polytomous key for the identification of species of the genus *Xiphinema* Cobb, 1913 (Nematoda: Longidoridae) with exclusion of the *X. americanum*-group: Supplement I. Systematic Parasitology 24:185-189.

Loof, P. A. A., M. Luc, and P. Baujard. 1996. A revised polytomous key for the identification of species of the genus *Xiphinema* Cobb, 1913 (Nematoda: Longidoridae) with exclusion of the *X. americanum*-group: Supplement 2. Systematic Parasitology 33:23-29.

Robbins, R. T., D. J. F. Brown, J. M. Halbrendt, and T. C. Vrain. 1996. Compendium of juvenile stages of *Xiphinema* species (Nematoda: Longidoridae). Russian Journal of Nematology 4:163-171.

Heteroderidae Filip'ev & Schuurmans Stekhoven, 1941

Mulvey, R. H., and A. M. Golden. 1983. An illustrated key to the cyst-forming genera and species of Heteroderidae in the western hemisphere with species morphometrics and distribution. Journal of Nematology 15: 1-59.

Wouts, W. M. 1985. Phylogenetic classification of the family Heteroderidae (Nematoda: Tylenchida). Systematic Parasitology 7:295-328.

Luc, M., A. Maggenti, and R. Fortuner. 1988. A reappraisal of Tylenchina (Nemata). 9. The family Heteroderidae Filip'ev & Schuurmans Stekhoven, 1941. Revue de Nématologie 11:159-176.

Heteroderinae Filip'ev & Schuurmans Stekhoven, 1941

Baldwin, J. G., and L. P. Schouest, Jr. 1990. Comparative detailed morphology of the Heteroderinae Filip'ev & Schuurmans Stekhoven, 1941, sensu Luc et al. (1988): Phylogenetic systematics and revised classification. Systematic Parasitology 15:81-106.

(Also, see key to genera of Heteroderinae by Baldwin and Mundo-Ocampo on pp. 345-346 of Manual of Agricultural Nematology)

This list of keys and references for identifying species of selected genera of plant-parasitic nematodes was compiled to assist students and professionals in the discipline of nematology. It is not comprehensive but was compiled as a current source of information and initial perspective for those interested in species identification of selected plant-parasitic nematode genera. Nomenclature and systematics follow that of Ebsary, 1991, and genera selected are primarily those in Mai, W. F., and P. G. Mullin, with H. H. Lyon and K. Loeffler. 1996. Plant-parasitic nematodes: A pictorial key to genera. Ithaca, NY: Cornell University Press. We added older references where most appropriate and included some references that are not keys, where keys are unavailable or if particularly useful as supplemental references. For information on older references and on genera not included herein, see E. C. Bernard and S. M. Baird. 1983. Bibliography of keys for the identification of plant-parasitic and free-living terrestrial nematodes. Nematology Newsletter 29(4):9-18. Detailed study should encompass original species descriptions. We recommend the following references for studies of the principles of taxonomy behind many of the keys and for general review:

Golden, A. M. 1971. Classification of the genera and higher categories of the order Tylenchida (Nematoda). Pp. 191-232 in B. M. Zuckerman, W. F. Mai, and R. A. Rohde, eds. Plant Parasitic Nematodes. Vol. 1. Morphology, Anatomy, Taxonomy and Ecology. New York: Academic Press.

The landmark collection of manuscripts entitled: A Reappraisal of Tylenchina (Nemata) published in Revue de Nématologie, 1987-88, by contributors R. Fortuner, E. Geraert, M. Luc, A. R. Maggenti, and D. J. Raski.

Siddiqi, M. R. 1985. Tylenchida parasites of plants and insects. Slough, UK: Commonwealth Agricultural Bureau on behalf of Commonwealth Institute of Parasitology. 645 p.

Ebsary, B. A., 1991. Catalog of the Order Tylenchida (Nematoda). Publication 1869/B. Ottawa, Canada: Canada Communication Group-Publishing. 196 p.

Nickle, W. R., ed. 1991. Manual of agricultural nematology. New York, NY: Marcel Dekker. 1035 p. (A valuable reference that has many species keys)

This list of references to the keys was prepared by:

S. A. Lewis, Department of Plant Pathology and Physiology, Clemson University, Clemson, SC 29634-0377

R. T. Robbins and Weimin Ye, Nematology Lab, Department of Plant Pathology, University of Arkansas, Fayetteville, AR 72701

Z. A. Handoo and A. M. Golden, USDA, ARS, Nematology Lab, Bldg. 011A, Rm. 159, BARC-West, 10300 Baltimore Avenue, Beltsville, MD 20705-2350

The authors thank Dr. E. C. Bernard for his review of this submission.

Appendix 7

Key to Females

1. Body greatly swollen; adult female immobile, completely or partially embedded in plant root.........2
1. Body relatively slender; mobile animals.........9

2. Tip of tail easily recognizable.........3
2. Tip of tail not recognizable; body round, oval, or bottle-shaped.........5

3. Body screw or spiral-shaped; head skeleton not developed.........4
3. Body C-shaped; head and skeleton well developed.........***Rotylenchulus***

4. Ventral excretory gland well developed.........***Trophonema***
4. Ventral excretory gland inconspicuous.........**Anguinidae**

5. Body shorter than 0.25 mm.........***Sphaeronema***
5. Body longer than 0.25 mm.........6

6. Body irregularly swollen; somewhat sausage- or kidney-shaped.........***Verutus***
6. Body regularly swollen, at most 3 times as long as wide.........7

7. Dead female forms a brown cyst.........**Heteroderidae**
7. Dead female does not form a cyst.........8

8. Excretory pore in front of median bulb.........**Meloidogynidae**
8. Excretory pore behind median bulb.........**Sphaeronematidae**

9. On the front side of the head, there is a mouth stylet or spear that can be projected....10
9. Head without mouth stylet or spear.........63

10. Mouth with tylenchid spear; esophagus usually with a median bulb.........11
10. Mouth with dorylaimid spear; esophagus cylindrical or bottle-shaped, without median bulb.........35

11. Median bulb without valve apparatus.........12
11. Median bulb with a valve apparatus.........17

12. Body extremely slender, ratio a greater than 60.........**Ecphyadophoridae**
12. Body less slender, ratio a smaller than 60.........13

13. Vulva more than two body diameters in front of anus..14
13. Vulva at most two body diameters in front of anus..15

14. Esophageal glands do not overlap intestine..**Tylenchidae**
14. Esophageal glands overlap intestine..***Pseudhalenchus***

15. Mouth spear curved; tail short, half-round. Gonads undeveloped...............***Dotylaphus***
15. Mouth spear not curved; tail cone-shaped...16

16. Lips are swollen; body longer than 2 mm..***Fungiotonchium***
16. Lips are not swollen; body shorter than 2 mm.....................................**Neotylenchidae**

17. Gradual transition between procorpus and median bulb, basal bulb strongly reduced; cuticle often conspicuously heavily ringed...18
17. Distinct constriction present between procorpus and median bulb.............................21

18. Cuticle not conspicuously coarsely ringed; constriction between median and basal bulb is relatively long...**Paratylenchidae**
18. Cuticle not conspicuously coarsely ringed, already visible at 40 x magnification; median bulb not much constricted, fused with the basal bulb...19

19. Cuticle consists of 2 layers, outer layer as-it-were forming an extra cuticle which surrounds the body like a sheath...20
19. Cuticle consisting of one layer; body plump and heavily ringed, rings often ornamented ....**Criconematidae**

20. Stylet knobs pointed forward; body plump, ratio a smaller than 20; vulva at over 90% (presumably of body length)..***Hemicriconemoides***
20. Stylet knobs rounded; body relatively slender, ratio a greater than 20; vulva at most at 90%...**Hemicycliophoridae**

21. One gonad...22
21. Two gonads..29

22. Lip area with four setae...**Atylenchidae**
22. Lip area without setae...23

23. Median bulb strongly muscular and conspicuously well developed; already conspicuous at low magnification; duct of dorsal esophageal gland opens into median bulb...24
23. Median bulb developed normally; duct of the dorsal esophageal gland opens behind the mouth stylet in the lumen of the esophagus...26

24. Esophageal glands do not overlap intestine..***Paraphelenchus***
24. Intestine dorsally overlapped by esophageal glands...25

25. Tail broadly rounded; stylet knobs lacking...*Aphelenchus*
25. Stylet knobs present; if absent, then tail not broadly rounded..........**Aphelenchoididae**

26. Head skeleton strongly developed; mouth spear conspicuously robust; tail usually shorter than 2.5 anal body diameters..**Pratylenchidae**
26. Head skeleton poorly developed; mouth spear slender; tail usually longer than 2.5 anal body diameters, rarely shorter..27

27. Tail relatively short, cone-shaped, oval; phasmids absent..........................**Anguinidae**
27. Tail drawn out long; rounded; phasmids at the level of the vulva, dorsal of the lateral field..28

28. Mouth spear long, longer than 40% of the distance from the front of the body to the median bulb..**Tylodoridae**
28. Mouth spear at most 30% of the distance from the front to the median bulb.................
....**Tylenchidae**

29. Oesophageal glands in the basal bulb overlap intestine..30
29. Oesophageal glands do not overlap intestine...32

30. Head skeleton strongly developed...31
30. Head skeleton not developed..**Dolichodoridae**

31. Lip region low, flattened at the front side...**Pratylenchidae**
31. Lip region high, round or flattened in front..**Hoplolaimidae**

32. Tail long; c' greater than 5; stylet knobs absent...**Psilenchidae**
32. Tail short, bluntly rounded; stylet with knobs..33

33. Head skeleton weakly developed..**Dolichodoridae**
33. Head skeleton strongly developed...34

34. Lateral field with 4 grooves...***Pararotylenchus***
34. Lateral field with 6 grooves...***Pararotylenchus***

35. Long, slender nematodes, greater than 2 mm; a ratio greater than 60 with a strongly elongated straight stylet, longer than 55 µm...**Longidoridae**
35. Stylet shorter than 55 µm, or when longer than the nematode is shorter than 2 mm; a ratio less than 50...36

36. Stylet relatively long and strongly bent; tail virtually absent.................**Trichodoridae**
36. Stylet straight; tail present..37

37. Two gonads...38
37. One gonad reduced..56

38. Oral cavity conspicuously large, wide, and sclerotized...39
38. Oral cavity inconspicuous..40

39. Oral cavity with 4 large teeth; in the wall of the oral cavity there is also a number of smaller teeth; tail long, head weakly set off, defined...............................***Paractinolaimus***
39. Oral cavity without large teeth; wall of oral cavity with 6 sclerotized ribs; tail short, head strongly set off...***Carcharolaimus***

40. Lip region set off by a deep constriction; lip region disc- or sucker-shaped; lateral field usually with a conspicuous series of glands...41
40. Lip region not disc- or sucker-shaped..42

41. Amphid opening located behind constriction..**Discolaimidae**
41. Amphid opening located in front of constriction...***Kochinema***

42. Stylet set in wall of oral cavity..43
42. Stylet axial...44

43. Stylet set in right subventral wall of oral cavity......................................**Nygolaimidae**
43. Stylet set in dorsal wall of oral cavity...***Sectonema***

44. Small, plump nematodes, with stylet made up of several components but often invisible because of grains on the body, with cuticle often loosely draped around the body ....**Diphtherophoridae**
44. Body without grains and with a normal cuticle..45

45. The posterior widened part of the esophagus is surrounded by a sheath of muscle..... ....**Belondiridae**
45. Esophagus proximally not surrounded by a muscle sheath......................................46

46. Stylet surrounded by a membranous guide ring; several layers of cuticle often easily discernible o the tail...**Aporcelaimid**ae
46. Guide ring not membranous; cuticular layers on the tail's tip not easily discernible... 47

47. Stylet asymmetrical, the dorsal side of the stylet is longer than the ventral side; stylet continuation is curved and surrounded by dense esophageal tissue............**Dorylaimoides**
47. Stylet, if asymmetrical, not with a stylet extension that is curved and surrounded by dense esophageal tissue..48

48. Proximal part of esophagus short and swollen, pear-shaped; transition of intestine to prerectum is situated close to the vulva..***Leptonchus***
48. Prerectum shorter, proximal part not swollen pear-shaped...49

49. Stylet extension with knob-shaped outgrowths; body length shorter than 1 mm; lip region set off like a cap...***Tylencholaimus***
49. Stylet extension without knob-shaped outgrowths, if with knob-stylet extension, then the lip region is not set off like a cap or the body length is greater..................................50

50. Cuticle with longitudinal ridges, large nematodes..................................**Dorylaimidae**
50. Cuticle without longitudinal ridges..51

51. Cuticle in oral cavity thickened, guide ring barrel-shaped...............**Chrysonematidae**
51. Guide ring not barrel-shaped, cuticle not thickened...52

52. Tail long, cone-shaped or finely drawn out......................................**Thornenematidae**
52. Tail short...53

53. Stylet conspicuously long 1/6 of the length of the esophagus..................***Longidorella***
53. Stylet at most as long as two lip diameters...54

54. Stylet continuation with a pear-shaped flange...***Enchodelus***
54. Stylet continuation, if flanged, not pear-shaped..55

55. Tail bluntly rounded, cylindrical; lip region not set off; stylet extension weakly flanged...***Thornia***
55. Tail not cylindrical, stylet extension not flanged.................................***Tylencholaimus***

56. Posterior gonad reduced..***Tylencholaimus***
56. Anterior gonad reduced...57

57. The wide proximal part of the esophagus is surrounded by a muscle sheath..............57
57. Esophagus not surrounded by a muscle sheath...59

58. Tail longer than 5 anal body diameters.......................................***Oxydirus oxycephalus***
58. Tail shorter than 1.5 anal body diameters..75

59. Stylet extension with knob-shaped outgrowths..60
59. Stylet extension without outgrowths..***Tylencholaimellus***

60. Stylet with dorsal reinforcement; lip region set off..78
60. Stylet without dorsal reinforcement; lip region not set off.............................***Dorylliun***

61. Stylet extension curbed and surrounded by dense esophageal tissue......***Dorylaimoides***
61. Stylet extension neither curved nor surrounded by dense tissue..................................62

62. Tail bluntly rounded; a few light-refracting particles around the mouth opening...***Pungentus***
62. Tail long or cone-shaped; no refractory particles around oral cavity...**Thornenematidae**

63. Body caterpillar-shaped with alternating wide and narrow rings; amphids bladder-shaped..**Desmoscolecidae**
63. Body not caterpillar-shaped...64

64. Body asymmetrical, tight body half displays a fine net structure and/or a series of fins or small shields...**Bunonematidae**
64. Body asymmetrical without conspicuous fins or small shields...................................65

65. Two gonads...66
65. One gonad..101

66. Esophagus with bulb, with valve apparatus..67
66. Esophagus without bulb, or, if swollen, without valve apparatus................................81

67. Bulb located halfway esophagus...68
67. Bulb at base of esophagus...70

68. Oral cavity tubular and with stylet knobs; esophagus tylenchid.........**Tylopharyngidae**
68. Oral cavity without stylet knobs..69

69. Oral cavity with a movable tooth dorsally; right subventrally a large tooth, and left subventrally a smooth or toothed small plate.......................................**Neodiplogasteridae**
69. Oral cavity with an immovable tooth dorsally; left and right metastoma swellings are identical to each other...**Diplogasteridae**

70. Cuticle punctate...71
70. Cuticle not punctate...75

71. Front of head with small crown..**Teratocephalidae**
71. Head without small crown...72

72. Front part of esophagus surrounding the oral cavity sharply set off by a constriction...
....**Ethmolaimidae**
72. Front part of esophagus not sharply set off..73

73. Amphids slit-shaped, inconspicuous...74
73. Amphids spiral-shaped, conspicuous..**Achromadoridae**

74. Oral cavity with a dorsal and several subventral teeth; distal part of esophagus symmetrical...**Chromadoridae**
74. Dorsal tooth is actually a continuation of the esophageal tissue; subventral teeth inconspicuous; distal part of esophagus asymmetrical.......................**Hypodontolaimidae**

75. Small crown on the front of the lip region......................................**Tetratocephalidae**
75. Lip region without crown..76

76. Valve apparatus divided by a transverse groove.....................................**Desmodoridae**
76. Valve apparatus not divided...77

77. Lumen of intestine elongated, conspicuous; amphid large and round...........................
....**Linhomoeidae**
77. Intestinal lumen not elongated, amphid not conspicuously large...............................78

78. Oral cavity uniformly tubular; tail glands lacking; amphids inconspicuous; head setae absent..79
78. Oral cavity narrowing at the base; tail glands and drainage duct present; amphid halfway oral cavity, round or slit-shaped; neck wings present or absent; head setae often present..**Plectidae**

79. Oral cavity not fused, consisting of separate rings; esophageal body with three consecutive swellings...23...**Alloionematidae**
79. Walls of oral cavity fused; body of esophagus at most regularly swollen..................80

80. Submedian lips curved like hooks...**Diploscapteridae**
80. Lips regular, not bent like hooks..**Rhabditidae**

81. Oral cavity large, barrel-shaped...82
81. Oral cavity small, elongate or absent...85

82. Oral cavity with one or more movable teeth...**Onchulidae**
82. Oral cavity without movable teeth..83

83. Tail drawn out like a thread; oral cavity small with barely sclerotized walls; head setae present...**Prismatolaimidae**
83. Tail never drawn out like a thread; oral cavity large and with strongly sclerotized walls; head setae absent..84

84. Base of oral cavity flat; base of esophagus with three small glands; usually three identical teeth in oral cavity..**Anatonchidae**
84. Base of oral cavity funnel-shaped; base of esophagus without glands; subventral tooth or teeth, if present, never identical to dorsal tooth......................................**Mononchidae**

85. Cuticle punctate..86
85. Cuticle not punctate..88

86. Amphids absent, tail bluntly rounded; oral cavity tubular, cuticle very finely punctate; ratio b greater than 10...**Isolaimidae**
86. Amphids spiral or funnel-shaped; ratio b smaller than 10..87

87. Tail short; ratio c greater than 30..**Choanolaimidae**
87. Tail long; ratio c at most 15...**Cyatholaimidae**

88. Base of esophagus swollen like a bulb, valve apparatus lacking................................89
88. Base of esophagus not swollen..92

89. Oral cavity absent; conspicuously large amphids...............................**Amphanolaimus**
89. Oral cavity present; amphids clearly smaller than half the corresponding body diameter...90

90. Oral cavity long, tubular; amphids inconspicuous. Setae absent.........**Rhabdolaimidae**
90. Oral cavity not tubular; amphids clear. Head setae present..91

91. Ratio c' greater than 6...**Odontolaimidae**
91. Ratio c' smaller than 6..**Diplopeltidae**

92. Oral cavity formed by a long thin tube which is at least 1/3 as long as the distance from the front to the esophageal base...**Aulolaimidae**
92. Oral cavity shorter...93

93. Amphid round or spiral-shaped...94
93. Amphid cup or chalice-shaped..97

94. Head with four setae; body conspicuously ringed; tail never spatula-shaped; behind the basal bulb the esophagus does not continue in the intestine..96
94. Head with 10 setae, 6 long ones and 4 shorter ones, or without setae; body not conspicuously ringed; tail spatula-shaped; behind the basal bulb the esophagus narrows and continues partially into the intestine esophagus with a basal bulb.........**Leptolaimidae**

95. Esophagus with a basal bulb...**Leptolaimidae**
95. Esophagus without basal bulb..**Halaphanolaimidae**

96. Oral cavity tubular with sclerotized walls..**Diplopeltidae**
96. Oral cavity inconspicuous, conspicuously slender nematodes.................**Bastianiidae**

97. Cuticle ringed; oral cavity closed at rest, but a tooth is recognizable in the closed oral cavity..**Tripylidae**
97. Cuticle smooth..98

98. Oral cavity tubular with three movable teeth in front......................................**Ironidae**
98. Oral cavity, if tubular, without movable teeth...99

99. Oral cavity tubular, with a small tooth halfway...................................**Bathyodontidae**
99. Oral cavity funnel-shaped or reduced..100

100. Head without setae; oral cavity inconspicuous..**Amphidelus**
100. Lip region with 10 setae; oral cavity large with a few little teeth...............**Tobrilidae**

101. Front gonad reduced...**Alaimidae**
101. Posterior gonad reduced...102

102. Anterior gonad flipped (turned around) and continuing to behind the vulva...........103
102. Anterior gonad, if flipped, not continuing to behind the vulva...............................105

103. Amphid conspicuously large, lip region set off, round..............................**Osstellidae**
103. Amphid inconspicuous...104

104. Anterior gonad continues to behind the vulva, where it bends sharply; cuticle usually coarsely ringed; oral cavity long and narrow, regularly sclerotized; cheilostoma usually wider than the other elements; edge of head often indented, lips with or without appendages...**Cephalobidae**
104. Anterior gonad continues beyond the vulva, and has no sharp bend; cuticle finely ringed; oral cavity consisting of cheilo-, pro-, and mesostome, which together form a more or less rectangular cavity; pro- and mesostome are strongly sclerotized; lips never with appendages ....**Panagrolaimidae**

105. Esophagus with bulb with valve apparatus..106
105. Esophagus without bulb; if weakly swollen at the base then without valve apparatus. 113

106. Bulb in the middle of the esophagus...**Diplogasteroididae**
106. Bulb at the base of the esophagus...107

107. Oral cavity tubular with parallel walls..**Rhabditida**
107. Oral cavity not tubular...108

108. On the front of the lip region there is a small crown...........................***Teratocephalus***
108. Lip region without crown...109

109. At the front of the head there are four long thin hairs; the bulb consists of two parts, in the anterior part a number of small teeth are present; the esophagus continues behind the posterior bulb into the intestine...***Chronogaster***
109. Head without long thin hairs; esophagus not continuing behind the posterior bulb...... 110

110. Body greater than 1 mm..111
110. Body smaller than 1 mm...112

111. Vulva at 55-62%...***Turbatrix***
111. Vulva at 70-80% ....***Plectonchus***

112. On the dorsal wall of the oral cavity there is a large tooth; anterior gonad turned around, head setae absent; amphid cup-shaped and inconspicuous....................***Campydora***
112. Oral cavity without teeth; gonad stretched out; head setae present; amphid conspicuous and round..***Monhystrella***

113. Oral cavity barrel-shaped...114
113. Oral cavity narrow...116

114. Esophagus consists of two parts, the anterior part is muscular, the posterior part consists of gland tissue; head setae absent; there are a number of small teeth in the oral cavity..**Odontopharyngidae**
114. The esophagus is not differentiated; head setae present; oral cavity at most with an inconspicuous tooth..115

115. Tail filamentous drawn out; amphids slit-shaped.............................**Prismatolaimidae**
115. Tail not drawn out filamentously; amphids round.............................**Sphaerolaimidae**

116. On the dorsal wall of the oral cavity there is a large tooth; head setae absent, amphid inconspicuous, cup-shaped...***Campydora***
116. Oral cavity without teeth; head setae present; amphid round and conspicuous.......117

117. Cuticle distinctly ringed; gonad located to the left of the intestine................**Xyalidae**
117. Cuticle smooth; gonad located to the right of the intestine..................**Monhysteridae**

Translated from Dutch by Dr. Anton Baudoin from the excellent resource written by Dr. Tom Bongers. (Used with permission.)

Bongers, T. 1988. De Nematoden van Nederland. Publication Foundation Royal Dutch Natural History Society, Library No. 46. Hoogwoud: Nederlands.

Appendix 8

Key to Males

1. Mouth stylet or spear present..2
1. Mouth stylet or spear absent..30

2. Tylenchid stylet present, esophagus usually with a median bulb with valve apparatus; ventral supplements lacking, a bursa is usually present..3
2. Dorylaimid spear present; esophagus usually bottle-shaped; a median bulb is lacking; supplements virtually always present, a bursa virtually always lacking.........................86

3. Lip region with four setae...**Atylenchidae**
3. Lip region without setae..4

4. Bursa absent..5
4. Bursa present..11

5. Median bulb conspicuously well developed; this is already conspicuous at low magnification. The duct of the dorsal esophageal gland opens into the median bulb; posterior part of body strongly curved ventrally...................................**Aphelenchoididae**
5. Median bulb not conspicuously strongly developed, posterior part of body at most weakly curved ventrally...6

6. Tail shorter than the anal body diameter...7
6. Tail longer than the anal body diameter...9

7. Body not twisted..***Verutus***
7. Body twisted...8

8. Amphid openings conspicuous...**Meloidogynidae**
8. Amphid openings on the front of the lip region, inconspicuous.............**Heteroderidae**

9. Esophagus differentiated, tail longer than 5 anal body diameters...............**Tylenchidae**
9. Esophagus degenerated, tail shorter than 5 anal body diameters..................................10

10. Median bulb still recognizable...***Trophonema***
10. Median bulb completely degenerated...**Paratylenchidae**

11. Bursa with rays...**Aphelenchidae**
11. Bursa without rays...12

12. Bursa adanal..13
12. Bursa surrounding tail..22

13. Body conspicuously slender, strongly reminiscent of a glass fiber; the body behind the anus is conspicuously narrower; bursa lobed.....................................**Ecphyadophoridae**
13. Body plump, ratio a smaller than 100...14

14. Esophagus and stylet degenerated..15
14. Esophagus and stylet not degenerated...16

15. Head skeleton weakly sclerotized...***Rotylenchulus***
15. Head skeleton not visible...**Paratylenchidae**

16. Median bulb strongly developed; duct of the dorsal esophageal gland opens just before the valve apparatus into the lumen of the esophagus; bursa rudimentary.................
....***Paraphelenchus***
16. Median bulb not conspicuously muscular, the bursa is well developed, the anterior part of the esophagus is swollen in the shape of a spool; a genuine median bulb is lacking.
....17

17. The anterior part of the esophagus is swollen in the shape of a spool; a genuine median bulb is lacking...**Neotylenchidae**
17. Median bulb well developed...18

18. Esophageal glands overlap the intestine..***Pseudhalenchus***
18. Esophageal glands do not overlap the intestine...19

19. Spicula widened at the front; tail cone-shaped, bursa reaches almost to the top of the tail..**Anguinidae**
19. Spicula not widened at the front; tail long drawn out; bursa adanal...........................20

20. Stylet knobs absent; phasmid on the tail; amphids slit-shaped, located at the side of the lip region..**Psilenchidae**
20. Stylet knobs present; phasmid at the middle of the body; amphids pore-shaped, inside view virtually invisible..21

21. Mouth spear long, reaching to 40% of the distance from the front of the lip region to the median bulb..**Tylodoridae**
21. Mouth spear shorter, at most to 1/3 of the distance to the median bulb......**Tylenchidae**

22. Esophageal glands overlap the intestine...23
22. Esophageal glands do not overlap the intestine..27

23. Median bulb and valve apparatus absent..***Deladenus***
23. Median bulb present...24

24. Head skeleton strongly developed...25
24. Head skeleton weakly developed 26

25. Lip region flattened at the front side..**Pratylenchidae**
25. Lip region at the front side hemispherical, body after heat fixation C-shaped...............
....**Hoplolaimidae**

26. Lip region dorsally move strongly developed than ventrally, spicule shorter than 20 mm...***Hoplotylus***
26. Lip region symmetrical, spicule longer than 20 mm...............................***Telotylenchus***

27. Esophagus without median bulb...***Paurodontus***
27. Median bulb and valve apparatus present..28

28. Head skeleton weakly developed..**Dolichodoridae**
28. Head skeleton strongly developed...29

29. Lateral field with four grooves...***Paraotyylenchus***
29. Lateral field with six grooves...***Amplimerlinius***

30. Esophagus with a conspicuous median bulb..31
30. Esophagus at most with a weak median bulb...33

31. Oral cavity tubular without conspicuous teeth..............................**Diplogasteroididae**
31. Oral cavity consisting of one or more sizable "rooms", teeth conspicuous................32

32. Dorsal tooth in the oral cavity immobile, the two subventral swellings of the metastoma are identical...**Diplogasteridae**
32. The dorsal tooth is mobile and protrusible, the right swelling of the metastoma contains a large tooth as well, the left swelling is smooth or carries a small toothed plate...**Neodiplogasteridae**

33. Body caterpillar-shaped, consisting of alternating wide and narrow rings; amphids bladder-shaped...**Desmoscolecidae**
33. Body not consisting of alternating wide and narrow rings..34

34. Body asymmetrical, the right half of the body displays a fine net structure and/or a series of small shields or fins...**Bunonematidae**
34. Body symmetrical..35

35. Bursa present..36
35. Bursa absent...42

36. Bursa without rays..37
36. Bursa with rays...38

37. Spicules with an elbow-shaped outgrowth, bursa lobed......................***Fungiotonchium***
37. Spicules without outgrowths..40

38. Spicules present, oral cavity tubular..39
38. Spicules absent, oral cavity with a massive dorsal tooth..............................***Myolaimus***

39. Submedian lips bent in the shape of a hook.......................................**Diploscapteridae**
39. Submedian lips not bent...**Rhabditidae**

40. Spicules C-shaped...**Hemicycliophoridae**
40. Spicules curved weakly...41

41. Body longer than 0.7 mm...***Loofia***
41. Body shorter than 0.7 mm...**Criconematidae**

42. Lip region with a small crown in front..**Teratocephalidae**
42. Lip region without crown..43

43. Esophagus degenerated and without structure...............................**Sphaeronematidae**
43. Esophagus well developed...44

44. Oral cavity tubular, with stylet knobs, tail drawn out like a thread
....**Tylopharyngidae**
44. Oral cavity without sytlet knobs, tail not drawn out like a thread..............................45

45. Amphids absent, oral cavity long and undifferentiated; large nematodes, longer than 3 mm, with a bluntly rounded tail and a conspicuously short esophagus, b ratio greater than 10...**Isolaimidae**
45. Amphids present, b ratio smaller than 10, tail not bluntly rounded and body shorter than 3 mm..46

46. Cuticle punctated, amphids usually spiral-shaped..47
46. Cuticle not punctated, amphids not clearly spiral-shaped supplements in the shape of a trapezoid, esophagus constricted at the front...53

47. Supplements in the shape of a trapezoid, esophagus constricted at the front...**Ethmolaimidae**
47. Supplements, if present, not in the shape of a trapezoid...48

48. Amphid opening slit-shaped..49
48. Amphid spiral-shaped..51

49. Cuticle not punctated...**Osstellidae**
49. Cuticle punctated...50

50. The dorsal and subventral teeth are equally large, the part of the esophagus surrounding the oral cavity is symmetrical...**Chromadoridae**
50. The dorsal tooth is small, it is in fact a continuation of the esophageal tissue; the subventral teeth are barely developed. The dorsal sector of the esophagus is wider than the ventral sector in front...**Hypodontolaimidae**

51. Tail blunt, cone-shaped and broadly rounded.....................................**Choanolaimidae**
51. Tail longer than 1.5 anal body diameters...52

52. Supplements absent..**Achromadoridae**
52. Supplements present..**Cyatholaimidae**

53. Supplements present..54
53. Supplements absent..68

54. Tubular supplements (tube-shaped)...56
54. Supplements not tube-shaped..55

55. Urn-shaped supplements...**Tobrilidae**
55. Supplements not urn-shaped..58

56. Anterior part of the valve apparatus with teeth.....................................**Leptolaimidae**
56. Valve apparatus, if present, without teeth...57

57. Bulb with valve apparatus. The esophagus does not continue into the intestines behind the posterior bulb...**Plectidae**
57. Bulb if present, without valve apparatus. The esophagus has a tongue-shaped extension behind the posterior bulb...**Leptolaimidae**

58. Oral cavity sizable, barrel-shaped...59
58. Oral cavity tubular or practically absent..62

59. Oral cavity with large movable teeth..**Onchulidae**
59. Oral cavity without movable teeth...60

60. Oral cavity small with an inconspicuous small tooth at the base. Tail drawn out like a thread..**Prismatolaimidae**
60. Oral cavity large with sclerotized walls and one or a few large teeth, tail never drawn out like a thread..61

61. Base of oral cavity funnel-shaped, the dorsal tooth is not identical to the subventral teeth; base of the esophagus contains no bladders..**Mononchidae**
61. The base of the oral cavity is flattened, the dorsal tooth is identical to the two subventral teeth; the base of the esophagus contains three bladders..............**Anatonchidae**

62. Oral cavity tube-shaped and longer than half of the actual esophagus.....**Aulolaimidae**
62. Oral cavity, if tube-shaped, shorter..63

63. Cuticle ringed..64
63. Cuticle smooth...66

64. Conspicuously slender nematodes, a ratio greater than 40.........................**Bastianiidae**
64. Body plump..65

65. Amphids inconspicuous, chalice-shaped. Oral cavity closed at rest, but a tooth is discernible in the oral cavity. Body spiral-shaped after heat fixation.................**Tripylidae**
65. Amphids round and conspicuous, oral cavity without teeth.....................**Diplopeltidae**

66. Oral cavity with several movable teeth...**Ironidae**
66. Oral cavity without movable teeth...67

67. Tail shorter than 2 anal body diameters..***Bathyodontus***
67. Tail long, body slender and so narrow that there is barely anything to be seen at the head, except for the amphids..**Alaimidae**

68. Posterior bulb with a distinct valve apparatus...69
68. Posterior bulb, if present without valve apparatus..79

69. Neck wings present...42...**Plectidae**
69. Neck wings absent..70

70. Valve apparatus divided in two by a cross groove..........................53...**Desmodoridae**
70. Valve apparatus without cross groove..71

71. Tail with tail glands and a terminal drainage duct 72
71. Tail glands and drainage duct absent..73

72. Setae absent; amphid inconspicuous; oral cavity without cross ridges...**Rhabdolaimidae**
72. Setae on the head and scattered over the body; amphids round; oral cavity with three cross ridges...**Linhomoeidae**

73. Lips with ciliated or forked lip appendages..**Cephalobidae**
73. Lips without lip appendages...74

74. Esophagus cylindrical..**Panagrolaimidae**
74. Esophagus swollen...75

75. Esophagus widens stepwise...**Alloionematidae**
75. Corpus cylindrical or spool-shaped...76

76. One of the spicules is shorter than the other...***Plectonchus***
76. Spicules of equal length..77

77. Metastoma with small teeth..**Panagrolaimidae**
77. Metastoma without teeth..78

78. Gubernaculum triangular with a caudal outgrowth.................................***Panagrobelus***
78. Gubernaculum slender, without caudal outgrowth.................................**Cephalobidae**

79. Head set-off..80
79. Head not set-off..81

80. Oral cavity small, inconspicuous; amphid cross oval.................................**Osstellidae**
80. Oral cavity large, egg-shaped; amphid round..................................**Sphaerolaimidae**

81. Amphids round...82
81. Amphids inconspicuous..84

82. Cuticle smooth...**Monhysteridae**
82. Cuticle finely ringed...83

83. Body longer than 1 mm...**Xyalidae**
83. Body smaller than 1 mm...***Domorganus***

84. Oral cavity large, with small teeth; right spicule degenerated......**Odontopharyngidae**
84. Oral small or absent, both spicules equally large...85

85. Body and head conspicuously slender (a ration greater than 60); amphid slit- or pore-shaped, at several body diameters behind the front of the body..........................**Alaimidae**
85. Body not conspicuously slender, amphid not discernible in side view..........................
....*Sphaeronema* or **Paratylenchidae**

86. A conspicuous suction-cup-shaped lip disc at the anterior side of the body..................
....**Discolaimidae**
86. No suction-cup-shaped outgrowth of the lip region..87

87. At the front end of the spear, there is a large oral cavity; on its base there are four large teeth and on the walls a large number of small teeth and 24 longitudinal ridges.........
....**Actinolaimidae**
87. Oral cavity with teeth lacking..88

88. Spear and other body content barely discernible since they are obscured......................
....**Diphtherophoridae**
88. Body content easily visible..89

89. Tail long; c ratio smaller than 15 and c' ratio greater than 2..90
89. Tail short; c ratio greater than 15 and c' less than 2..95

90. Basal part of the esophagus surrounded by a muscular sheath.........................*Oxydirus*
90. Esophagus not surrounded by a muscular sheath..91

91. Nine or more supplements...92
91. Eight or fewer supplements..93

92. Tail cone-shaped; four anal body diameters long...................................*Eudorylaimus*
92. Tail drawn out like a thread...*Prodorylaimus*

93. Stylet continuation curved and surrounded by swollen esophageal tissue.....................
....*Dorylaimoides*
93. Stylet continuation not surrounded by swollen esophageal tissue.............................94

94. Cuticle in the oral cavity swollen in the shape of a barrel, c ratio greater than 9...........
....*Chrysonemoides*
94. Cuticle in the oral cavity not swollen in the shape of a barrel, c ratio less than 9..........
....*Prodorylaimium*

95. Stylet and stylet extension fused into a curved spear; in front of the anus there are three supplements, there are a few ventral papillae in the esophageal region. Bursa sometimes present...**Trichodoridae**
95. Spear straight, no ventral papillae in esophageal region...96

96. Spear longer than 70 µm..**Longidoridae**
96. Spear shorter than 70 µm..97

97. Body longer than 3 mm..98
97. Body shorter than 3 mm..101

98. Spear is inserted into the ventral wall of the oral cavity, is dorsally grooved, and with a basal outgrowth..***Sectonema***
98. Spear axial..99

99. Cuticle with longitudinal ridges; 35-45 supplements..***Dorylaimus***
99. Cuticle without longitudinal ridges; fewer than 20 supplements..............................100

100. Spicule longer than 80 µm; spear opening 30% of the spear length..........***Axonchium***
100. Spicule shorter than 70 µm; spear opening 50%; base of esophagus not surrounded by a muscular sheath..***Aporcelaimus***

101. Stylet continuation with knob-shaped outgrowths................................**Leptonchidae**
101. Stylet continuation without knobs..102

102. Spear curved like a sickle body conspicuously narrowed at the anterior side; body width at the base of the esophagus is 4.5 times the width at the lips; guide ring membranous, folded; twenty supplements of which one is isolated by itself..***Paraxonchium***
102. Body not conspicuously narrowed..103

103. Stylet continuation with pear-shaped flanges..***Enchodelus***
103. Stylet continuation without flanges..104

104. Oral cavity covered by six membranes..***Torumanawa***
104. Oral cavity not covered by membranes..105

105. Spear embedded/inserted in right subventral wall of the oral cavity; base of esophagus surrounded by a muscular sheath; supplements poorly developed..**Nygolaimidae**
105. Spear axial..106

106. Supplements, gubernaculum, and lateral guide pieces lacking.......................***Thornia***
106. Supplements present..107

107. Stylet continuation curved; esophageal tissue swollen behind the spear..**Dorylaimoides**
107. Stylet continuation not curved; esophagus not swollen distally.............................108

108. Guide ring membranous; spear opening greater than 50%; gubernaculum absent; three separate cuticular layers easily discernible..**Aporcelaimidae**
108. Tail cone-shaped, rounded or pointed, never broadly rounded...............................109

109. Tail cone-shaped, rounded or pointed, never broadly rounded........**Qudsianematidae**
109. Tail bluntly rounded...110

110. Oral cavity shielded by six small shields; 19-21 grouped supplements; body longer than 2 mm...***Labronema***
110. Oral cavity without shields..111

111. Tail with a number of cuticular granules; lip region not set off; guidering consisting of a muscular collar...***Thonus***
111. Tail without granules...**Thornematidae**

Translated from Dutch by Dr. Anton Baudoin from the excellent resource written by Dr. Tom Bongers. (Used with permission.)

Bongers, T. 1988. De Nematoden van Nederland. Publication Foundation Royal Dutch Natural History Society, Library No. 46. Hoogwoud: Nederlands.

Appendix 9

Formulas for Fixatives and Stains

Acid fuchsin lactophenol

Phenol (liquid)....................50 ml
Lactic acid....................50 ml
Glycerin....................100 ml
Distilled water....................50 ml

To prepare stain, add 5 ml of a 1% stock solution of acid fuchsin per 100 ml of lactophenol. The concentration of stain may need to be changed, depending on the age of tissues.

FA 4:1

Formalin (40% formaldehyde)....................10 ml
Glacial acetic acid....................1 ml
Distilled water....................top up to 100 ml

FA 4:10

Formalin (40% formaldehyde)....................10 ml
Glacial acetic acid....................10 ml
Distilled water....................80 ml

FAA

95% ethanol....................20 ml
Formalin (40% formaldehyde)....................6 ml
Glacial acetic acid....................1 ml
Distilled water....................40 ml

Formalin (2%)

Formalin (40% fromaldhyde)....................5 ml
Glycerol....................2 ml
Distilled water....................60 ml

FP 4:1

Formalin (40% formaldehyde)....................10 ml
Propionic acid....................1 ml
Distilled water....................89 ml

Glutaraldehyde fixative (2%)

Sodium cacodylate buffer....................33 ml
70% glutaraldehyde....................2 ml

Hechler's Solution

Formalin (40% formaldehyde)....................4 ml
Glycerol....................6 ml
Distilled water....................90 ml

Hydeman's Plant Fixative

Ethyl alcohol....................33 ml
Distilled water....................24 ml
Fomalin (40% formaldehyde)....................3 ml
Glycerol....................3 ml

Lactophenol

Liquid phenol....................20 parts
Lactic acid (45%)....................20 parts
Glycerine....................40 parts
Distilled water....................20 parts

Propionic-orcein

Orcein stain (natural or synthetic)....................2.2 g
Glacial propionic acid....................100.0 ml
Boiled gently for 20 minutes (Warning: boiling can suddenly become violent)
Cool the solution to room temperature
Diluted with distilled water....................100 ml

Seinhorst I

Ethanol (95%)....................20 parts
Glycerine ....1 part
Distilled water....................79 parts

Seinhorst II

Ethanol (95%)....................95 parts
Glycerine ....5 parts

Sodium cacodylate buffer 0.1 M

Sodium cacodylate....................21.4 gm
Distilled water....................500 ml

TAF

Formalin (40% formaldehyde)....................7 ml
Triethanolaimine....................2 ml
Distilled water....................91 ml

Note: All solutions are very toxic and should be handled only with protective gloves, goggles, etc. and under a fume hood.

Made in the USA
Monee, IL
02 November 2026

67381621R00122